D0122548

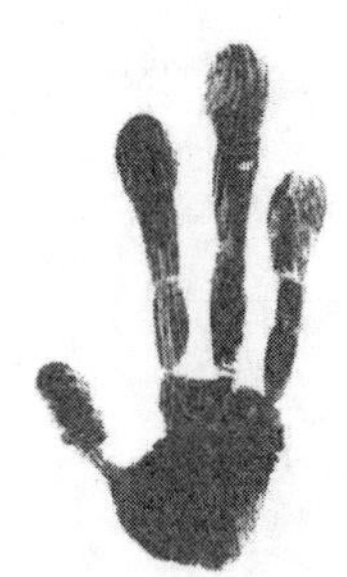

CONFESSIONS *of* AN ALIEN HUNTER

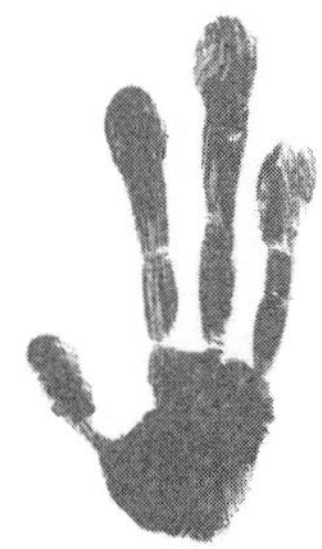

A SCIENTIST'S SEARCH FOR EXTRATERRESTRIAL INTELLIGENCE

SETH SHOSTAK

Library of Congress Cataloging-in-Publication Data
Shostak, G. Seth.
Confessions of an alien hunter : a scientist's search for extraterrestrial intelligence / by Seth Shostak.
p. cm.
ISBN 978-1-4262-0392-3
1. Life on other planets. I. Title.
QB54.S549 2008
576.8'39--dc22
2008046731

Founded in 1888, the National Geographic Society is one of the largest nonprofit scientific and educational organizations in the world. It reaches more than 285 million people worldwide each month through its official journal, *National Geographic*, and its four other magazines; the National Geographic Channel; television documentaries; radio programs; films; books; videos and DVDs; maps; and interactive media. National Geographic has funded more than 8,000 scientific research projects and supports an education program combating geographic illiteracy.

For more information, please call 1-800-NGS LINE
(647-5463) or write to the following address:

National Geographic Society
1145 17th Street N.W.
Washington, D.C. 20036-4688 U.S.A.

Visit us online at www.nationalgeographic.com

For information about special discounts for bulk purchases, please contact
National Geographic Books Special Sales: ngspecsales@ngs.org

For rights or permissions inquiries, please contact National Geographic Books
Subsidiary Rights: ngbookrights@ngs.org

Interior design: Cameron Zotter

Printed in USA

CONTENTS

FOREWORD

The trajectory of human history has frequently been changed by major events which redirected that path of history for good: for example, the rise of major religions, and the impact of major wars. In modern times, examples are the rise of Hitler and communism, and the development of antibiotics and nuclear energy. Sometimes these events required time to develop, as in the case of the industrial revolution, with the eventual consequences not recognized as the event transpired.

This book explores such a major world-changing event, which is in progress right now. Its eventual impact is largely unknown and thus largely unappreciated, although intuition tells us that the impact will be enormous and all to the good. It is the discovery of life in outer space, and of intelligent life in particular.

The idea of life in outer space is a very old one, going back to the Greek philosophers. The existence of other inhabitable worlds and extraterrestrial life has for hundreds of years been one of the few very prime questions in science. Not only would their discovery provide a bonanza of new scientific information, it would provide great insights concerning the significance and destiny of humanity: Where did we come from, are we unique, what is our place in the cosmos, and what might our future be?

The first major steps in learning of extraterrestrial life have occurred in little more than the last decade. Most important has been the detection of not just one, or a few, but hundreds of other planetary systems, some containing planets resembling the Earth, although not yet near twins to the Earth—those are undetectable with current telescopes. History will judge this development to be one of the greatest milestones in the history of science. Yes, the best is yet to come. In the next few years, new instruments, such as the Kepler spacecraft, should tell us of the existence and abundance of Earthlike planets. This will usher in a wave of great discoveries about the abundance and detailed nature of Earthlike planets.

For more than 40 years, searches for signs of extraterrestrial technology, a surrogate for extraterrestrial intelligent creatures, have been taking place using the instruments of optical and radio astronomy to search for radio and light signals from other worlds. The rate of improvement in these systems is astonishing and will continue, perhaps even at an increased rate.

All of this progress has been the product of a limited group of ingenious and dedicated people, a characteristic typical of previous major forks in the path of history.

Seth Shostak has been fortunate to be an active participant in this group as it grew from a very small size to the large cohort of today. His description here of this history, and the remarkable people involved, comes directly from much firsthand personal experience, and can be counted on to be true.

Searching for extraterrestrial signals is one of the most challenging tasks ever taken on by mankind. There are more than a hundred billion stars which might, according to our limited present knowledge, be the source of such signals. The radio and optical spectrum allows the existence of perhaps billions of signal channels. We are faced with exploring a multitude of combinations of stars and channels for an elusive signal, whose actual form we can only guess. We are challenged to use logic to predict what another civilization, probably much older and more advanced than us, might adopt as a technology we might detect. Here we have no choice but to enter a realm of ill-informed speculation about the abilities and motivations of advanced civilizations. To reach an answer, we have to become futurists, reaching far beyond our usual comfortable world of telescope technology to arrive at possible scenarios for the distant future. This becomes an exercise of intellect reaching far beyond the usual bounds of science theory. It is the provocative theme of the latter part of this book. Is the future of sapient creatures to become symbionts with computers of unimaginable capability? To become, in a way, immortal, through our machines, and to be intelligent beyond our wildest dreams? If so, what will the signals be? What will the messages be? Will we even be able to recognize them, let along understand them? History is not encouraging—our signals of today are very different from the signals of 40 years ago, which we then felt were perfect models of what might be radiated from other worlds of any state of advancement. We were wrong. If technology can change that much in 40 years, how much might it change in thousands or millions of years? Our powers of prediction so far have not been up to the task of answering this question.

So we must assume that the search will be long and demanding of time and resources, both material and human. We will be groping in the dark. Realism demands that we accept this scenario. But to Seth Shostak, and the others who join him in the search, it is all worth it because the end result will be discoveries leading us along a new and incredibly rich fork in the long road of terrestrial history.

—Frank Drake

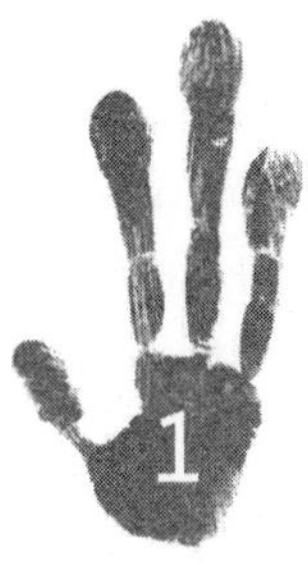

NEWS THAT WOULD CHANGE THE WORLD

It felt as if I was sweating on the inside. I was anxious and nervous, confronting a disturbing, if melodramatic, thought: This might be the most important day in the history of *Homo sapiens.*

The clock was easing past 3 a.m. on June 24, 1997, and we hadn't left the institute. On most nights, this room would be forlorn, just another bit of office space in the Silicon Valley, passively waiting for morning like a garaged car. But tonight the ceiling fluorescents hummed and flickered, while a half dozen co-workers sat in a line like Easter Island *maoi,* gazing at computer monitors. All that could be seen on the screens were a few lines of text and a graphic display. But they had our full attention.

Hour after hour the hushed vigil continued, while on the far side of the country a radio telescope paced a dim star across the sky. Hundreds of millions of people were asleep in the landscapes between this room and that telescope, and they knew nothing of this. Not yet. But there was a chance—and we thought it a good chance—they would soon face a world that had changed forever.

I looked again at a small, luminous pane of gray dots on the screen in front of me. To one side was a faint lineup of brighter dots, a white slash across the gray. That was the signal; that was the cosmic static mesmerizing us at this wee hour. There wasn't any sound (we had never bothered to hook up audio equipment to the computers), only the slowly marching phalanx of white dots. This signal, which we had picked up early the previous morning, had galvanized my colleagues at the radio telescope in West Virginia. Now, nearly a day later, we were watching it by remote control in California.

The possibility of discovery was making me fidgety. My immediate worry, ironically, was about trivial matters. If the signal turned out to be real—if it was really beamed from aliens—then my schedule for the week was going to be completely messed up. Everything would need to be canceled or postponed—every appointment, every meeting, every lunch and dinner. It also occurred to me that this disruption was

Monitoring the candidate signal at 3:30 a.m., June 24, 1997

going to persist into the following week. And the week after that. I couldn't sit down. I kept pacing the length of the office, occasionally taking photos as a justifiable diversion. At no point did it occur to me, or to any of the others in the room, that maybe someone should go out for food. Or just go home and sleep.

This was too important. We were on the verge of proving that humankind had company, that other intelligence dwelled among the stars. But I wondered if we were ready to hear from beings that would make *Homo sapiens* look like an also-ran.

If you grab the next dozen people off the street and ask them if they believe intelligent beings inhabit worlds beyond our own, the chances are good that more than half will smile at your question and then nod affirmatively. A 2005 poll commissioned by the National Geographic Channel and conducted by the University of Connecticut's Center for Survey Research and Analysis found that 60 percent of Americans believe that "life exists on other planets."

The poll question was vague, of course, because "life" might mean no more than the type of microscopic, single-celled entities that, despite their widespread prevalence, get short shrift in most people's minds (unless they're either pathologists or in the yeast business). But many people presumably interpreted the survey question as an inquiry about *intelligent* life—the type of complex, self-aware beings that routinely show up in science fiction stories to either befriend or beleaguer us Earthlings.

Although pollsters haven't specifically asked scientists, I suspect that the majority would also agree with this premise, confiding their own belief that sentient extraterrestrial life exists. Most of the literati I encounter are convinced that "someone must be out there." Of course, my academic colleagues perhaps are being kind, taking pains not to offend my delicate sensibilities by suggesting that the search for other intelligence among the stars is futile. Then again, maybe they feel a need to demonstrate their appreciation of astronomical history, to stand with Copernicus and Galileo against the Earth-centered cosmology of Aristotle. For the last half millennium, it's been both fashionable and a pleasant demonstration of modesty to decry any suggestion that our world is special.

Be that as it may, there's good and growing reason to expect cosmic company. The past hundred years have witnessed a slow but inexorable spring tide of discoveries that encourage the idea that Earth may have many, many analogues—even within the provincial confines of our home galaxy. The prospects even in our local neighborhood are more promising than they used to be. We once reckoned that among the hundreds of planets and moons of the solar system, only the Earth and Mars were endowed with the conditions that would permit life to endure (and we weren't sure about Mars). However, thanks to reconnaissance by many tens of spacecraft and a handful of landers, we've recently identified a half dozen other members of the Sun's household that might have liquid water, and therefore should be added to our list of habitable locales.

That list of wet worlds is not restricted to planets. In the late 1990s, NASA's Galileo spacecraft detected changing magnetic fields in the vicinities of Europa, Ganymede, and Callisto—three of Jupiter's largest moons. Magnetic fields can result from electric currents in salty water, and the Galileo data imply that massive oceans hide beneath the thick, frozen skins of these satellites. In principle, such perpetually dark, saline seas could have spawned an indigenous flora and fauna, although it's unlikely they would contain anything as sophisticated as a squid or a tuna.

Even frigid Titan, Saturn's largest moon, might be home to biological life. NASA's Cassini spacecraft has recently mapped lakes on the surface of this smoggy world. But these are not lakes of water. A thermometer on the surface of this forlorn moon would read -290°F (-179°C)—a temperature so abysmal that water freezes like rock. However, beneath Titan's gelid epidermis lies a reservoir of ammonia-laced water that—on occasion—might burst forth in a sort of low-temperature volcano. Recent studies hint that the brief mixing of this regurgitated liquid with surface hydrocarbons might foment the type of chemistry that underpins life.

We still don't know if these worlds—or any of the other possible locales for life in our solar system—are truly host to biological activity, but our concept of habitable worlds has become much broader in just a few decades. Life is surely wondrous, and, for most of us, the most interesting activity in a universe filled with interesting activity. The fact that we've uncovered habitats for biology that were thoroughly unanticipated only a generation ago suggests a scenario in which life is not just decorative scrollwork on the cosmic edifice but part and parcel of its basic structure. Life may be sacred, but it might also be commonplace.

WORLDS AROUND OTHER STARS

Although our space programs have revealed unexpected venues for life nearby, ground-based researchers have made an equally remarkable discovery: They've found new solar systems in our galactic backyard. In 1995, two Swiss astronomers accidentally tripped over evidence of a Brobdingnagian planet that is frenetically orbiting a rather ordinary star 50 light-years away. Since then, hundreds of additional worlds have been found. We now suspect that the *majority* of stars have planets. This is, without a doubt, one of the most dramatic astronomical discoveries of the past two decades.

We still don't know how many of these distant orbs will prove to be the type of worlds where complex, cogitating life might rear its brainy head. Many of those we've discovered so far are "hot Jupiters"—big planets in sizzling, star-hugging orbits. But recent results from these same Swiss researchers and others suggest—to no one's surprise—that small planets will turn out to be more plentiful than large ones. This is pleasant news because jumbo worlds are often swathed in smelly, thick atmospheres of methane, ammonia, and other heavy gases. These ingredients are more suitable for fueling buses or making fertilizer than powering biology. Small planets, on the other hand, are the sort that might be wrapped in the thin atmospheres and watery seas that seem so evidently suited to life as we know it. Small worlds have a chance of being Earthlike worlds.

That the universe contains an abundance of planets seems certain, but does that prove that living things—big-brained or pea-brained—are out there? Of course it doesn't. But that caution should be balanced with a remarkable fact: Astronomers haven't found any reason not to believe life exists in space, even though *it could have been otherwise.* They might have learned that planets were uncommon, or that worlds where oceans could exist were rarer than steak tartare. The former is simply untrue, and recent research strongly suggests that the latter is quite unlikely.

In addition, planets are no longer the only qualifying real estate that could sustain life. Large moons, once regarded as mere planetary accessories, are now acknowledged to be perfectly legitimate candidates for habitation.

So the universe promises to have many niches where life could begin and evolve. A rough estimate is that more than a hundred billion such planets and moons pepper our own galaxy. If that's an insufficient number to encourage your faith in cosmic company, note that nearly a hundred billion *other* galaxies are visible to our telescopes.

Not only have we learned that the cosmos could be rife with life, we've bettered our techniques for finding it. Much of that improvement derives from our ability to travel to other worlds. In 1920, a *New York Times* editorial writer ridiculed the idea of rockets sailing through space, saying that such craft would fail to work because the engine would lack "something better than a vacuum against which to react." Less than a half century after that uninformed remark, men brought by rockets were driving golf balls on the moon.

Today, we shoot hardware and the occasional human skyward at Mach 35, or more than ten times the speed of a rifle bullet. That's fast enough to reach the moon in days, Mars in months, and Pluto in a decade. These rockets are enabling us to uncover life—if it exists—on the idiosyncratic worlds of our solar system in the most straightforward manner possible: by finding it in situ. Such a discovery could happen at any time—possibly even before your next dental appointment.

But given the brutal conditions on these nearby moons and planets, any life we uncover in our solar system won't be any bigger—or brighter—than a paramecium. If our goal is to find intelligent aliens—the type that can reason as well as (or better than) we can—we'll have to search much farther afield.

We won't be doing that in person. The rockets that routinely slip the bounds of Earth run on high-test fuels such as kerosene or hydrazine. That gives them enough zip for excursions within the solar system. But chemical rockets are far too pokey to take us or our robotic proxies to the stars—the presumed locales of other advanced, intelligent beings. Our fastest spacecraft would take 25,000 years to traverse a single light-year. Suppose, in a fit of bravado, we were to launch a probe seeking clever aliens in one of the star systems of Orion's belt. Assuming all went well and the craft successfully found and photographed a planet carpeted by extraterrestrials, the news wouldn't reach our descendants (assuming there were any) until the year 20,000,000.

ANOTHER APPROACH

That's a tedious wait. But nearly a half century ago, two scientists at Cornell University realized that it was feasible to send signals between star systems at the speed of light. Radio waves are especially practical in this regard, because they're easy to generate, require relatively little power, and slice right through any intervening gas and dust. Indeed, radio is such an efficient method of interstellar telegraphy that many astronomers have come to the conclusion that, whatever else might occupy the minds of extraterrestrials, they'll surely be emitting radio signals for space communications.

This immediately led the Cornell scientists to urge astronomers who study the natural radio static from space to use their antennas to hunt for *artificially* produced interstellar broadcasts. Quick and simple calculations had shown that our best radio receivers are sensitive enough to pick up our most powerful transmitters at enormous remove—the distances of the stars. If we could hear ourselves from light-years away, then presumably we could hear advanced aliens. This was equivalent to a realization along the lines of "Hey, our wooden ships are good enough to cross oceans. Maybe we should look for new continents."

It was a remarkable idea whose brilliance has been dulled by long familiarity: Hardware that we've already built for other purposes, accessible right here on the ground, could discover sentient beings living on unseen worlds. We could find cosmic company with no more effort than turning a dial.

Of course, scientists can cite manifold reasons why a reconnaissance for alien signals, an enterprise now known as SETI, the Search for Extraterrestrial Intelligence, might come up empty. Even aside from the many technical uncertainties (where on the dial would the aliens broadcast? At what power level? Would it be a continuous radio whine, or just an occasional "ping"?), there are many ways that SETI could fail no matter what level of effort is expended in its pursuit. For example, while many biologists are comfortable with the premise that life will commonly spring up in suitable habitats, they will wrinkle their forehead at the suggestion that life-encrusted worlds will inevitably, or even occasionally, produce intelligent beings. As evolutionary biologist Stephen Jay Gould took pleasure in noting, *Homo sapiens* is no more than an evolutionary fluke, and if we could "rewind the tape of life" and play it again with slightly different circumstances, neither you nor Gould would ever have trod the planet. His point of view is not shared by all biologists, but it resonates with more than a few.

Other researchers note that advanced aliens could long ago have colonized every star system in the galaxy—the universe has afforded more than enough time for such a venture.

And yet our solar system doesn't seem to be part of any interstellar imperium, a discordant circumstance known as the Fermi paradox (physicist Enrico Fermi pointed out this conflict between expectation and observation in 1950). For many people, this is strong evidence that no one else clever enough to build rockets or radio transmitters calls our galaxy home. They conclude that we're the smartest species within a few hundred thousand light-years or more, a thought that should simultaneously flatter and dismay you.

These considerations, together with others, provide ample fodder for those who maintain that any search for sentient

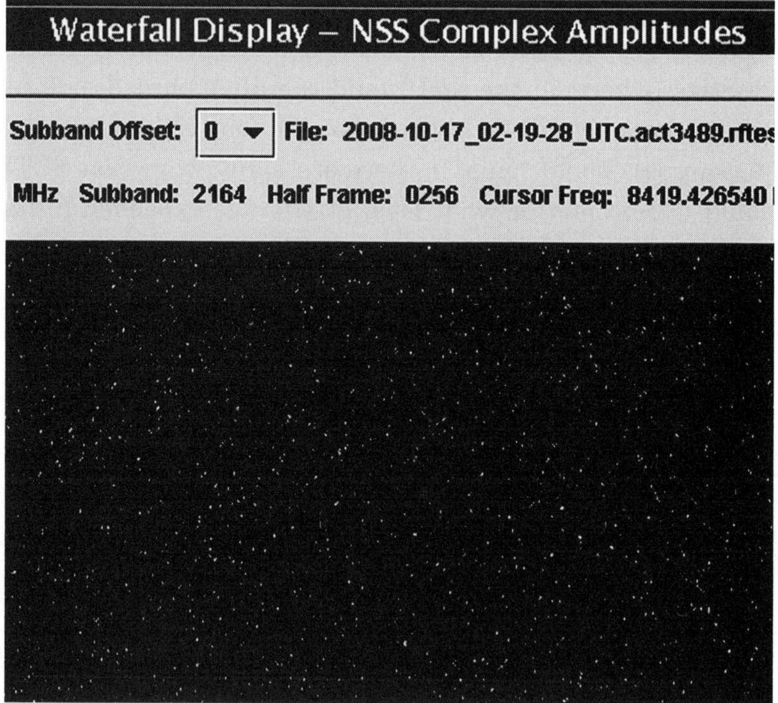

A radio signal from the distant Voyager 1 spacecraft (faint line at right) is representative of what a true SETI signal would look like.

life beyond Earth is quixotic. But in the end, these arguments are akin to banter about the existence of giant squids. The classical Greeks talked about them, as did medieval mariners. But were these supersize cephalopods mythical or real? The discussion was settled when parts of the creature were actually snatched from the seas by a French gunboat 150 years ago.

Debate is entertaining, but exploration and experiment are definitive. So SETI researchers have decided to tune out the arguments and try to tune in the extraterrestrials. They've also steadily improved the tools they use. In 1960, astronomer Frank Drake conducted the first modern SETI search using an antenna smaller than a baseball infield and a receiver that could monitor one radio channel at a time. He searched for signals from two nearby stars. By the late 1990s, the most sensitive SETI experiments boasted a custom-built receiver able to monitor 56 million channels simultaneously, and were conducted on large antennas such as the 140-foot radio telescope in Green Bank, West Virginia, and the Arecibo radio telescope in Puerto Rico. The latter has a metal dish that could easily swallow the entire Rose Bowl.

It was during SETI observations using the 140-foot Green Bank antenna that the signal of June 24, 1997, was found.

The small crowd staring at the computer monitors shifted restlessly in their chairs. It was now nearly 4 a.m., and the star at which our antenna was doggedly staring—YZ Cet, in the constellation Cetus (the Sea Monster)—continued its climb above the southeast horizon, dragged upward by the slow spin of Earth. YZ Cet was, as far as we knew, not an especially promising neighborhood for sentient life. A dwarf flare star, smaller and dimmer than the Sun, it was prone to periodic hot flashes that would be discomfiting to any planets in close orbit. But YZ Cet had one overriding good quality: It was only 12 light-years away. Our present search, Project Phoenix, was scrutinizing all stars within 15 light-years, no matter what their character. And YZ Cet might have planets; who knew? Flares or no, one of those planets might have technically sophisticated beings. A transmission from this star system was possible.

The 140-foot antenna swiveled smoothly on its massive ball-bearing mount, keeping the star in its radio sights. At the telescope, directing the observations, was Jill Tarter, the SETI Institute's lead researcher in the hunt for alien signals and the prototype for the Ellie Arroway character in astronomer and author Carl Sagan's novel (and film) *Contact.* Early the previous morning, Tarter had welcomed a television crew to the observatory. They had come to shoot a segment for yet another documentary about extraterrestrial life. Thanks to the endless popularity of aliens, such TV crews are not infrequent, and in my experience, when they arrive at various telescopes, they inevitably believe that a message from the stars will arrive as well. It's an appealing arrogance. We've been searching for years without success, but these guys figure that on the one day they show up for a photo op, so will an alien signal.

Ironically, it seemed possible that this crew might, indeed, have hit the bull's-eye. With the scientists buzzing about the signal from YZ Cet, the crew switched on their Betacams and started burning tape at warp speed. No one in the control room had time to talk to them.

A long 22 hours had elapsed since the radio bleep had been found. Clearly, it was far louder than the confusing background surf of receiver noise. But what sort of signal was it? We could pick from only two plausible choices: It was either a false alarm—static from a man-made transmitter causing a useless adrenaline rush—or it was the real deal. No middle ground.

Despite the roiling confusion, we tried to figure out what we were measuring. For a modern SETI experiment, signals are like graffiti—commonplace and mostly uninteresting. This antenna, this hulking metal mirror, was outfitted with a 56-million-channel receiver. Every few minutes it examined another swath of the radio spectrum, angling for the jackpot. With that many channels, each of these short observations was the statistical equivalent of pulling the handle on a Las Vegas slot machine day and night for 30 years.

If we couldn't rule out an extraterrestrial origin and turn this signal into just another false alarm—if we couldn't prove that this was merely the hand of man—Tarter would be under increasing pressure to call up another observatory and sweet-talk them into double-checking our find. Of course, the call could be embarrassing if the frenzy turned out to be caused by nothing more than unrecognized interference, equipment malfunction, or a software bug. The SETI enterprise is regarded by most astronomers with respect: It's scientific, it's serious, and it's diligent. But if you break into someone's observing run with an urgent request to stop what they're doing and check out a suspected alien signal, well . . . that's getting astronomers where they live. They're busy doing research that might lead either to some breakthrough in their field or, maybe, to tenure. While any astronomer would be pleased to take part in a truly momentous discovery, they'd be less sympathetic to a disruptive wolf cry. Nonetheless, Tarter had already begun scanning her list of observatories, noting those that might be able to certify our signal.

The previous day, the SETI team in West Virginia had tried all the usual tests for discriminating human broadcasts from those made by extraterrestrials. They were hampered in this effort by an unusual problem. An auxiliary telescope in Georgia that would normally be used to check suspected signals was on the sidelines, taken out of action by a broken axle bearing. It wouldn't be repaired for weeks.

So the Green Bank team made do. They tried swinging the 140-foot antenna to and fro, between the star's position and a sector of sky that was nearly seven degrees away. The signal alternately appeared and vanished, suggesting that its source was fixed at one spot on the sky. They did this repeatedly, proving that the spot moved with the stars as the Earth turned. This is a rigorous test, and few signals pass. Telecommunications satellites—a frequent source of interference—don't spin with Earth's rotation. They're either at low altitude and dance across the sky in minutes, or else they're in geosynchronous orbit (your favorite television satellite service, for instance), hanging fixed and forever at one position.

John Dreher, a supple-minded physicist working alongside Tarter, decided on a second test. Rather than shift the antenna quite so far off YZ Cet's position, he wanted to try nodding it a mere 0.1 degree away, the width of a knitting needle held at arm's length. Much as a searchlight beam falls off in intensity near its edges, so too does the sensitivity of an antenna decrease when a transmitting target is not dead center. At 0.1 degree, any signal coming from YZ Cet should be reduced in strength by half. If the radio dimming was actually observed, Dreher reasoned, then he had yet another strong argument that the signal was coming from somewhere near the dwarf star. Presumably a planet. An inhabited planet.

The test required that special commands be sent to the SETI Institute software driving the antenna. That code had largely been written by Jane Jordan, a software engineer back in California. So Dreher rang up Jordan at 6 a.m. California time and asked her to arrange for the new test. Jordan set this up using some backdoor code in her software that allowed her to reprogram the telescope from her home computer. The test was made, and the results were . . . odd. The signal was just as strong 0.1 degree away from YZ Cet as dead on.

"I announced these facts at our stand-up meeting at the institute that morning," Jordan said years later. "But people either didn't understand or didn't listen. Some continued to think that this star was still a candidate, possibly ET's broadcast."

Indeed they did. Tarter, who regularly worked the night shift, had been planning to fly back to California that morning, turning the observing over to others. But the drama of YZ Cet was too compelling. She left a 5 a.m. voice mail for her assistant, Christine Neller, asking that her flight be rescheduled. Tarter was determined to stay and check out the signal. Meanwhile, John Dreher had begun to suspect that the bright line of dots lined up on the computer screens might be caused not by ET, but by a European research satellite.

On the West Coast, not all of what was happening at the telescope was either obvious or comprehended. At about 6 a.m., YZ Cet dropped below the wooded West Virginia hills flanking the observatory to the southwest. It would be out of sight for 16 hours. In California, the sleepy crowd pushed back their chairs and leaked out of the room. I briefly considered finding a couch in the upstairs office and getting some sleep. But my mind was still whirling with questions. In particular, what if this signal kept passing the tests? At what point were we going to let others know?

In some sense, this last question was already irrelevant. There's no secrecy in SETI, and I was sure that everyone in the room had already spoken to many people about what we had found. Indeed, I was surprised that no one from outside had yet called or otherwise shown interest. It brought to mind a recent television interview in San Francisco where I had been paired with Don Ecker, editor of *UFO Magazine*. The first question the host posed to me was what would happen if SETI found a signal. I responded by maundering on about verification, notification, and similar niceties.

Ecker, clearly agitated, broke in by claiming that, if we found a signal, the military would show up at the observatory door and shut us down.

Shut us down? I didn't believe Ecker then, and I certainly didn't believe him now. Nobody was shutting us down. Heck, the mayor of Mountain View, the Silicon Valley city that's home to the SETI Institute, didn't call, and I knew him personally. Neither did any other local officials, let alone the Feds.

But what were we going to say? Should we hold a press conference?

That was one possibility. Full disclosure. But another possibility was to follow the lead of the Cambridge University team that had discovered pulsars, the enormously regular cosmic clocks that accidentally came to light in 1967. For months the British scientists kept back the news of these tantalizing ticks because they weren't sure what they were.

The research staff referred to the strange emitters as LGMs (little green men) in a backhanded reference to the possibility—the possibility—that they might be signals from an alien civilization.

Ascribing a greenish hue to the putative broadcasters was tongue in cheek, of course, but the idea that this was an artificially produced signal was not. No one had yet thought of a way by which a star might emit such a precisely repeated train of radio pulses. Only after the astronomers had found two more flashing radio sources located elsewhere in the sky, thus strongly indicating a widespread and natural origin, did the Cambridge team go public.

Maybe this was the right strategy: Keep mum until we had proof, one way or the other. But philosophically I disliked this approach. My long experience with UFO believers had made me acutely sensitive to their claim that SETI would keep secrets. I was opposed to keeping secrets, but on the other hand, a premature announcement that we'd found ET followed by later analysis showing we'd made a mistake would surely tarnish our enterprise. We didn't want to cry wolf, but we also didn't want to keep silent if the wolf was wandering the halls.

I was back at my desk shortly after nine o'clock when the phone rang. It was William J. Broad, science writer for the *New York Times*.

"So, Seth," he said, "what about that signal you're following?" My first thought was, how did he know? But I didn't ask that, and I didn't consider pretending that we weren't following a signal.

"Well, we're continuing to track the star," I replied. "But you know, these things often turn out to be man-made interference. We're checking out a lead on that right now." I waited for a reaction, and when there wasn't any, I continued: "But I think we'll know more in three hours or so. Can I call you back then?" Broad assented.

As I hung up the phone, I was suddenly aware of a dead-obvious insight. Everyone who had given any thought to how an extraterrestrial signal would be detected always assumed it would be like in the movies. Some nerdy-looking, bored SETI scientist would be monitoring a receiver when suddenly the signal would pop up on the screen. The scientist would shout "Eureka!" like Archimedes stepping into the bathtub. The people who accepted this scenario—and they included the SETI scientists themselves—planned on press conferences, alerts to the astronomical community, and other such punctilious procedures on the basis of a sudden revelation, a jump from "no, no, no" to a delirium-inducing "yes!"

Clearly, it wasn't going to happen that way. More than 24 hours had elapsed since we had first twisted the 140-foot antenna toward YZ Cet, and we still didn't know whether to shout "Eureka!" or "Nuts!"

Within two hours, the jury returned to the courtroom. The evidence was overwhelming and the verdict was clear: This was not ET. One tip-off was the bizarre 0.1-degree test—inconsistent with a transmitter at the position of YZ Cet. But the clincher was the fact that the transmission perfectly matched the telemetry signal from SOHO—the Solar and Heliospheric Observatory—a joint European and American mission to study the Sun from space. This billion-dollar satellite had been launched in late 1995 and was orbiting slowly a million miles from Earth, in the Sun's direction. A ten-watt onboard transmitter radioed SOHO's data back to NASA's antennas, and, as it happened, to ours as well. By chance, the way the signal bounced around the steelwork of the 140-foot had mimicked the behavior of an extraterrestrial source. If that second telescope down in Georgia had been operational, we wouldn't have been fooled.

Before lunch, I called back the *New York Times* and told Bill Broad that the signal was man-made. Much later, I figured out how he learned of the detection. Broad had been working on a biographical story of Carl Sagan, who had died six months earlier. As part of that effort he called Sagan's widow, Ann Druyan. Her secretary had already been in touch with Christine Neller, looking for Jill Tarter. The secretary was told that Tarter was still in West Virginia, postponing her flight to check out a signal. That innocuous tidbit was passed to Druyan, who passed it to Broad, who called me. There truly is no secrecy in SETI.

The finding was disappointing, but the discouragement was temporary. After all, there were hundreds of billions of other galactic stars we hadn't examined. In addition, the search was bound to get faster because of the steadily expanding power of computers.

I was also buoyed by a recent dramatic series of discoveries. Astronomers had just learned that planets were commonplace, and space scientists had found other possible habitats lurking in our solar system—watery locales that were unsuspected previously. I was suffused with the feeling that a hunt for life beyond Earth was now an enterprise with a real chance of success. In the history of humankind, this was something new.

INKLINGS OF ET

Today, the idea of space aliens—intelligent beings inhabiting far-off worlds—is as familiar as frozen food. This is partly the result of speculation by scientists. But the major reason for thinking that aliens roam the universe is their slimy presence in modern popular culture. Books, comics, movies, video games, and television are all liberally seasoned with creatures from other star systems. In many cases, these imaginary extraterrestrials have left their native worlds to invade ours—a good scenario from the storyteller's point of view, as it both simplifies the setting and amplifies the horror. You may not be losing sleep over the internecine struggles of aliens hundreds of light-years away, but smooth-skinned invaders on a mission to methodically obliterate *Homo sapiens* and trash our planet will attract your notice.

Today, fictional aliens are so ubiquitous they should probably unionize. Yet the idea that extraterrestrial societies stud the cosmos traces its roots back further than any sci-fi fantasy. Once our simian precursors were clever enough to be aware of their own existence, they must surely have gazed

The 1955 film This Island Earth *featured anthropomorphic aliens with big heads, big eyes, and brains on the outside.*

on the night sky, upholstered with its mysterious, luminous geography, and wondered, "Is someone like us up there?"

The classical Greeks thought they knew the answer, mainly because they had a theory regarding the deep structure of the universe. While usually remembered for inventing high school math, we should also credit the Greeks with being first in print with the idea of "other worlds." To them, the cosmos consisted of everything they could see: the Sun, planets, and naked-eye stars. But they also had the temerity to suggest that there might be *other* universes entirely separate from what they could espy on heaven's dark vault.

By definition, these other cosmos could not be observed or measured. The argument for their existence sprang from the uncomplicated atomic theory that Democritus, Epicurus, and others espoused around 400 B.C. According to these celebrated Mediterranean gentlemen, our world was created by swarming atoms that collided and coalesced after dropping into the local neighborhood from an infinite netherworld. But since the number of atoms was presumed to be infinite, and as only a finite quantity of atoms was required to construct the visible cosmos, these early Hellenic philosophers concluded that an unlimited number of parallel cosmos must be out there.* In an admirable act of economy, the Greeks took pragmatic advantage of the "spaces" between these cosmos to provide habitats for their gods.

Of course, postulating other suns and moons is one thing, but should we also expect them to host life? The poet Lucretius argued yes in his six-volume tome *De Rerum Natura (On the Nature of the Universe)*. Lucretius debunked the superstitious hogwash rampant in Roman religious practices, writing in passionate hexameter about the need "to rend the crossbars at the gates of Nature old." On the matter of cosmic habitation, Lucretius drew an obvious conclusion from the Greek atomic theory: If there really was an infinite

---

**This virtually unprovable philosophical notion has a modern-day counterpart. Some theoreticians have speculated that our own universe is but one of many parallel, noninteracting universes—the so-called multiverse hypothesis.*

amount of raw material in the form of atoms, they would produce anything that *could* be produced, including other living creatures.

By the Middle Ages, Lucretius's expansive view on the possibility of extraterrestrial life had taken a backseat to the more conservative outlook of one of his predecessors, Aristotle. Aristotle challenged the multiworld notion of the atomists, insisting that the observed cosmos was unique. He argued that the Earth was the center of the universe and the sole home of thinking beings. The antithetical views of Aristotle and Lucretius presented medieval society with a philosophical conundrum. Did they wish to subscribe to the invigorating notion of cosmic company, or prefer the pride (and loneliness) of being the universe's top—and only—dog?

For most of a millennium, the Western world opted for the latter, especially after Aristotle's point of view was written into Catholic theology by St. Thomas Aquinas. The idea of a unique and central Earth attained the status of conventional wisdom.

The Aristotelian worldview not only accorded well with the medieval desire to reconcile natural law and Scripture but also meshed nicely with the cosmology Ptolemy advocated in the second century A.D. As every schoolchild is told, Ptolemy postulated an Earth-centered universe, with the moon, Sun, planets, and stars situated on concentric, crystalline spheres. The universe was an immense, perfectly constructed clockwork onion, and Earth was at the center.

Not, however, at *dead* center. The planets were assumed to move in circles, but anyone who pays the slightest attention will note that they traverse the sky with variable speed. The Ptolemaic model was modified to account for this nonuniform behavior by the addition of geometric gimcracks akin to wheels within wheels. As a scheme for predicting the positions of the planets, this model worked, in much the same way that a Rube Goldberg contraption works. But it was complicated conceptually.

Renaissance superstar and sometime cleric Nicolaus Copernicus successfully assaulted Ptolemy's celestial citadel in 1543 when he published *De Revolutionibus Orbium Coelestium (On the Revolution of the Celestial Orbs).* Driven by aesthetic and philosophical considerations, Copernicus suggested replacing Ptolemy's Earth-centered system with a cosmology in which the planets revolve around the Sun.

De Revolutionibus hit the bookshelves barely half a century before Columbus's discovery of the New World. It is difficult to say which event more radically altered the minds of Europe. Copernicus had shifted the center of the cosmos to the Sun—a mere 93 million miles (150 million kilometers)—a trifling distance in astronomy. And yet the philosophical consequences were profound. With a Sun-centered universe, the Earth had become simply another heavenly body, and therefore no more privileged than Venus, Mars, or Jupiter.

So the possibility of cosmic confreres was back on the table: If our planet's astronomical situation wasn't unique, perhaps its locus as a home to intelligent life wasn't either.

As the 17th century got under way, some Renaissance academics began to go beyond mere speculation about extraterrestrials—they began looking for them, using a new instrument that did a remarkable thing: It brought distant views closer.

MEN ON THE MOON

The first telescopes—no more than simple lenses mounted at opposite ends of a tube—revealed that the moon, contrary to theological belief, had a less-than-perfect complexion. It was disfigured by an assortment of lumps and bumps, features that from a distance were reminiscent of earthly landscapes. Telescopes also disclosed that the bright, perambulating points of light known as planets were vaguely like Earth. At least, they were round.

As these first telescopic scrutinies of the sky revealed one celestial secret after another, the idea that we could actually *find* aliens, or at least uncover indirect evidence for their existence, took hold. Galileo Galilei was the first to make a systematic examination of the moon, espying things that 10,000 previous generations of moon-watchers had never seen. He noticed the peculiar circular depressions that we now call craters, and he also found mountains, which were easily discerned as they brightened in the first hours of a lunar morning. The length of their shadows revealed that some were higher than any of the Apennine *montagnes* gracing Galileo's Florentine neighborhood. However, despite the moon's apparent similarity to Earth (he likened one rugged region to Bohemia), Galileo didn't succumb to easy temptation and claim that our rocky, natural satellite was inhabited. In fact, he was doubtful it was.

A contemporary, Johannes Kepler, was less cautious. Enthralled by Galileo's work, Kepler made his own telescopic reconnaissance of the moon. In addition to the mountain chains, he saw deep valleys (now called rilles), which he assumed were cut by running water. For Kepler, this distant landscape appeared familiar and suggestive, and his imagination clicked into overdrive. He postulated that this nearest of worlds was crawling with humanlike beings. Since the scale of the lunar topography was grander than Earth's, with higher peaks and lower valleys, Kepler concluded that the moon men were probably taller than we are (a bit of loopy logic that, if applied on

Earth, might encourage the Nepalese in their basketball ambitions). He remarked on an apparent absence of cities, but argued that these were buried beneath the aforementioned craters. Kepler reasoned that, since lunar daylight lasts for two weeks at a time, subterranean cities would be more comfortable.

Kepler was arguably the first person to offer observational evidence for aliens. But his claims began to wither 150 years later when a mathematician by the name of Roger Boscovich pointed out a few disturbing obstacles to the moon's suitability for life. The shadows of the mountains were deep black, not gray, as would be expected if substantial quantities of air were present. Also, when the moon slid in front of distant stars, they blinked out suddenly without any dimming or reddening. The English astronomer William Herschel, who noted this effect during an eclipse in 1837, concluded that the air on the moon had to be at least 2,000 times thinner than on Earth. Three decades later, William Huggins tried to measure any lunar atmosphere by looking for changes in the colors of the yellow-orange star Epsilon Piscium when it ducked behind the moon. He saw no effects whatsoever, thereby confirming Herschel's grim assessment of breathing opportunities for lunarians. By the mid-19th century, astronomers knew that the moon was a suffocating place.*

If these facts didn't quash speculation about lunar life, the absence of any liquid water should have. The moon had no clouds, no snow, no flashes of sunlight on lakes or oceans. The moon was drier than a Mark Twain quip.

**Of course, the moon has some atmosphere, but not much. Measures made during the Apollo missions indicated that the total weight of all the moon's air was no more than 100 tons. This is the same amount stacked up on a patch of Earth the size of a pickup truck.*

Consequently, by the late 19th century, only the terminally naive were still hoping to find life on the moon. This was not a completely empty category. In 1875, the *New York Times* ran a story in which Russian astronomers at the Pamlateska Observatory claimed to see a bright orange light coming from the moon's edge. According to the Russians, this was a signal from backside burghers trying to get our attention by using a large mirror, 100 feet (30 meters) in size or more, to reflect sunlight back to Earth.

This was an interesting story (if nuttier than peanut brittle), because it postulated a situation in which aliens—nearby aliens—were trying to grab our attention. Kepler's claims had been based on passive and indirect evidence; he inferred the presence of moon dwellers from the topography of their world. In contrast, the extraterrestrials suggested by the Russians were taking the trouble to initiate contact.

The idea of sending messages between worlds was not new. It had been tentatively explored nearly four decades earlier when a Scottish clergyman, Thomas Dick, laid out the case for interworld communication in a book with the charmingly alliterative title *Celestial Scenery*. After pointing out that many eminent astronomers still thought the moon was inhabited, Dick regretfully noted that direct proof would be hard to come by. Even if the moon's denizens were the size of elephants, our best telescopes couldn't see them as they strutted the leaden landscape of their dimpled world. But, Dick said, if they had cities (aboveground) or other large-scale civil engineering projects comparable in size to the largest on Earth—those *could* be found.

Dick hatched a plan for discovering this evidence. A hundred observers would peer through their telescopes for 30 or 40 years, meticulously scrutinizing this nearby world (Dick did not address the matter of funding). He also reprised an idea that had been broached a decade earlier by the German mathematician Carl Friedrich Gauss, namely breaking the ice with the aliens by mowing giant geometric figures (triangles or ellipses) into Siberian grasslands. These clearly artificial large-scale constructions could be picked out by lunar astronomers with *their* telescopes. Dick argued that signaling otherworldly neighbors was surely a more admirable enterprise than the "millions which are now wasting in the pursuits of mad ambition and destructive warfare." As it turned out, neither of these projects was actually undertaken, although one could argue that they would have constituted the first major SETI project.

After Huggins, the obsession with moon men waned, gullible Russians excepted. Our natural satellite was obviously, clearly, and irrefutably as dead as dandruff. Attention shifted to Mars, a world that—unlike the moon—seemed to have both weather and water. And, perhaps, Martians.

These putative Martians were in trouble, though. Their world was drying out, and only a planetwide civil engineering project could keep them from droughty doom. The well-worn story of the Martian canals—a misbegotten product of the Victorian era—was in full swing by the end of the 19th century.

MARTIAN PUBLIC WORKS

In 1877, Giovanni Schiaparelli, director of Milan's Brera Observatory, created provocative charts showing the red planet was webbed by dark lines that were straight as a skewer. For three decades, he toyed with the idea that these "*canali*" were part of an advanced civilization's attempts to address a chronic water shortage. Canali, after all, could refer either to natural channels or deliberately dug water courses. Schiaparelli's own opinion shifted back and forth. Sometimes he admitted the unusual features might be mere topography. At other times he seemed convinced they were the product of spade-wielding Martians. It's undoubtedly relevant that Schiaparelli had studied hydrology at school.

Most other astronomers failed to see the straight lines, but Schiaparelli's story resonated with a young American, Percival Lowell—a man whose imposing presence, protean intellect, and unbridled self-confidence ensured that the story of the canals would forever be bonded to his name.

Lowell was the scion of wealthy Boston parentage: one branch of a family that had spurred the cotton mills that were the warm-up act for America's industrial revolution. In the factories of the early 19th century, it should be noted, water power spun the wheels and turbines. Lowell learned of Schiaparelli's reports about Mars and, intrigued by the Italian's claims of extraterrestrial trench work, decided to study the matter more deeply. To do so, he needed an observatory. After sending scouts around the globe to determine the best location for examining the sky, he settled on Flagstaff, Arizona. In 1894, on a hill at the edge of town, Percival dedicated his new dome, modestly naming it the Lowell Observatory.

Almost immediately, the enthusiastic Bostonian was busy charting the canals of Mars, features that he had unabashedly anticipated and that were obligingly easy to see. By 1895 he published the first of several books discussing these perfectly direct and remarkably long structures (he estimated their average length as more than a thousand miles). Lowell found 183 canals, tracing their undeviating paths on white, soccer ball–size globes. He insisted that the canals had been constructed deliberately. To him, that was as obvious as vaudeville humor. Lowell's facility with language and engaging lecture style ensured that his books, and his arguments, found wide sympathy among the public. At the Lowell Observatory today, he is cited as "the Carl Sagan of his time." Soon everyone knew of the canals, even if not all were convinced of their reality.

Lowell's special ability to discern planetwide civil engineering wasn't confined to Mars. He also claimed to see linear features on Venus's bright white physiognomy. Here the canals resembled spokes, radiating from a near-central

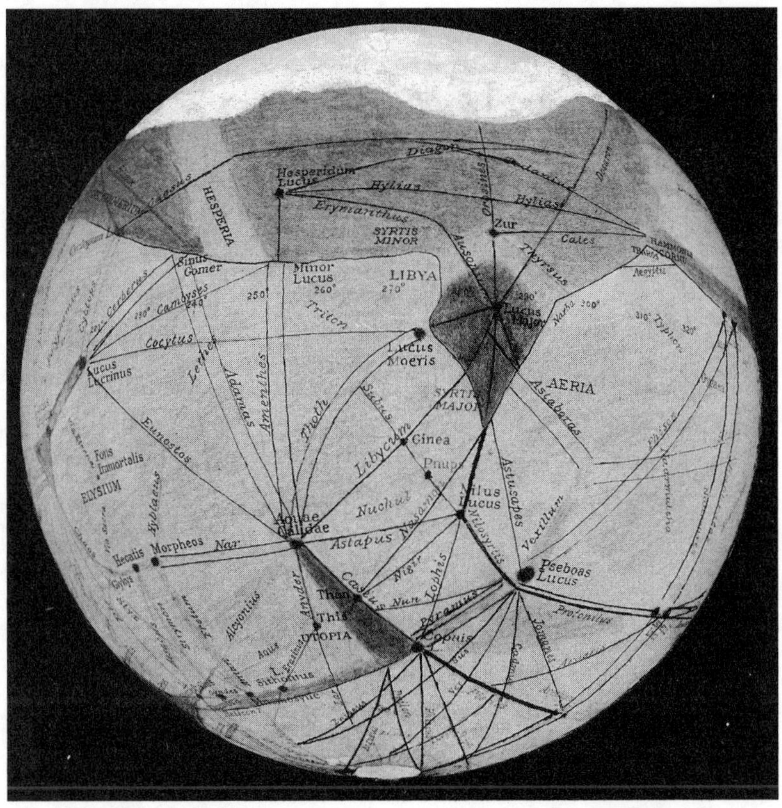

Percival Lowell's 1905 drawing of Mars, showing canals

point. There seemed to be a lot of ditch digging going on in the solar system.

Lowell died in 1916, convinced to the last that the Martian canals were real (he had become less vocal about Venus). By then, few astronomers shared his view. They had turned far larger telescopes than Lowell's on the red planet and failed to see the fine network of features that had caused all the commotion. A 1903 experiment conducted by the English solar astronomer Edward Maunder offered an explanation for the claimed canals. Maunder showed British schoolboys small disks on which various "features" had been drawn mimicking the Martian topography (sans canals). He found that at a certain distance, the boys would mentally merge small details on the disks into straight features. Their brains were evidently wired to connect the dots. Maunder suggested that the canals were optical illusions, not fluvial features on a faraway world.

LAST-DITCH EFFORTS

The thoughtful reader will wonder why the canal controversy wasn't settled by photography. Even in the 1890s, this art was already more than a half century old. So why didn't Lowell or his detractors decide the canal controversy by taking some pictures? Admittedly, astronomical photography was still primitive in the 1890s, but at least the camera would have been an impartial observer.

The difficulty was that it would also have been a *crippled* observer. Filling the column between the 18-inch objective lens of Lowell's telescope and the inert vacuum of outer space was one and a half tons of air. Air is turbulent, churned into waves and eddies by the terrestrial day-night cycle and by jet streams. These irregularities and motions cause stars to twinkle, which, while poetically useful, is disastrous for astronomy. Most of the time Lowell would have seen Mars dancing like a water bead on a hot skillet—especially at high magnification. Observing Mars was like mapping the design on a dime at a distance of 330 feet (100 meters), and doing so through more than a ton of simmering air. Any photos would be abysmally blurred.

Nonetheless, Lowell did encourage a few efforts to verify the canals' existence using cameras, and in 1907 he financed a photographic expedition to Chile by Amherst College astronomer David Todd. But the pictures taken by Todd were equivocal, and visual observations remained the bulwark of Lowell's argument. In a cold, dark dome, Lowell continued to sketch the evidence for Martian canals.

As the 20th century ticked over, the astronomical community grew increasingly dismissive of Lowell's claims. Nonetheless, the canal story had soaked into the collective unconscious, and there was no wicking it back out. A few people still hoped for Martians, and the Roaring Twenties begat a new opportunity to justify their optimism.

In August 1924, Mars was in opposition—not in a strategic sense, but in an astronomical one. It was opposite the Sun in the sky, placing the red planet substantially closer to Earth than usual—indeed, closer than it had been since 1804—and offering a particularly auspicious opportunity to get in touch. By this time a tempting new method of communication had been discovered.

Radio was the high-tech product of the day. Theoretically predicted in the 1860s, radio waves were first produced on a lab bench by Heinrich Hertz in the late 1880s. Within two decades, Guglielmo Marconi had improved radio apparatus to the point that he could send a telegraph signal across the Atlantic.

With the development of the vacuum tube and the lifting of restrictions on broadcasting after the First World War, radio mutated from scientific curiosity to household commodity. As the 1920s dawned, the Westinghouse Corporation quickly put four AM transmitters on the air, including KDKA in Pittsburgh—the first commercial radio station in the world. By 1923, approximately 500 AM stations were broadcasting in the U.S., prompting an event grandly known as International Radio Week. Its goal was to demonstrate the feasibility of routine transatlantic radio communication by encouraging listeners in America and England to alternately tune in one another's stations for a half hour every night at 10:00. The country on the receiving end was supposed to turn off its transmitters, while those in the transmitting country let loose with all the power their tubes would tolerate. The result of this "can you hear me now" test was that some of the stations were heard some of the time.

Despite the lukewarm results, the public was enthused by the idea of long-distance radio. The proponents of sentient Martian life quickly understood that if radio communication between continents was feasible, surely they could send signals between worlds. Perhaps an advanced culture on the red planet had beefy transmitters and would aim them our way during the 1924 opposition in a kind of interplanetary reprise of International Radio Week.

An attempt to tune in the Martians was organized by David Todd, the same astronomer whom Lowell had dispatched to Chile in 1907. Todd still believed that the canals were real, and that signals from Mars might be wafting our way. He requested that every commercial and military radio transmitter in America go off the air for five minutes each hour during a two-day period, although virtually none did. When astronomical research and a radio station's profits were at loggerheads, there seems to have been precious little agonizing over what to do. Nonetheless, several receiving stations noticed unusual broadcasts. The excitement was considerable; it was also short-lived. Signals picked up in Vancouver, British Columbia, were eventually traced to Washington State, where officials were developing a new radio navigation system for inland waterways (this, at least, had some connection with canals). Another turned out to be a commercial broadcast from Louisville, Kentucky. Several so-called whistlers were produced when the ionosphere perverted the natural static of lightning bolts into a radio signal that shifted in frequency like a slide whistle.

Throughout the 1924 "big listen," the Army's top cryptologist, William Friedman (who would unravel Imperial Japan's top-secret Purple Code two decades later), stood at the ready. After all, if a Martian message was received, it might not be in plain text. Alas, no interplanetary signal was received and Friedman was left buffing his nails. The clod-busting, canal-crazed Martians remained coy. Mainly because there weren't any.

THE REAL MARS

That sophisticated Martians were in short supply became indisputable four decades later when we became able, for the first time in the history of humankind, to observe the red planet up close. NASA's Mariner 4 spacecraft spent a half year bridging the featureless space between Earth and Mars, and in June 1965 it quickly flew past its target, coming within

6,000 miles (10,000 kilometers) of the stony terrain below. Sharing the view, it dutifully sent 22 black-and-white digital photographs back to California at an unimpressive one byte per second. The pictures were low resolution but high impact. Here, finally and unambiguously, was the real Mars: littered with craters and riven by dusty canyons. Its atmosphere was as thin as a dollar bill, devoid of much oxygen or nitrogen. The vast hydraulic society that many believed peopled this planet was an illusion.

Admittedly, by 1965 not many academics still thought the canal story held water. But Mars's desertlike landscape was dispiriting, and the scientists lowered their expectations. Big-brained Martians weren't in the cards, but there was still a sporting chance for some less imposing forms of life. Perhaps lichenlike life was clumped between the rocks, or lizardlike creatures were burrowing in the sand. An embryology professor at the California Institute of Technology, Albert Tyler, cheerfully opined that finding life on the red planet necessitated no more than a mousetrap and a camera. At the very least, microbes might be discovered in the soil. Martian life was still taken seriously enough to prompt one of the most sophisticated space reconnaissances ever ventured—the Viking mission.

In a remarkable technological tour de force, NASA set two spacecraft down on Mars's rufous surface. The first of the Viking landers arrived on July 20, 1976. Six weeks later so did its twin (on the other side of the planet). The landers' first task was to activate TV cameras to give humans their first ever close-ups of the Martian scenery.

Today, such photos are as familiar as your brother's eating habits. But in 1976, to behold this reddish, rocky terrain was revolutionary. As the Viking cameras opened their shutters,

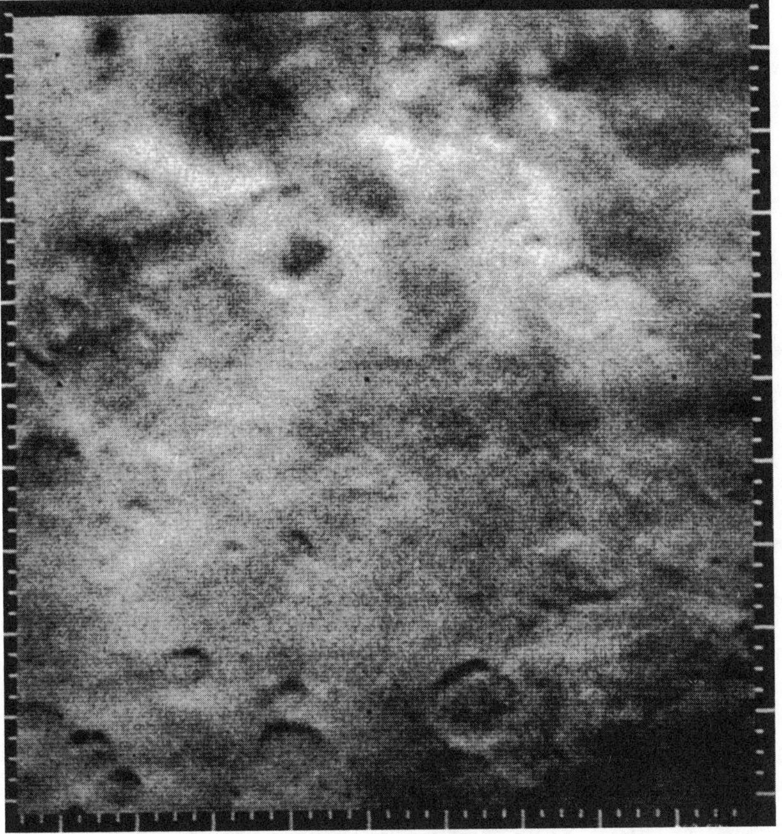

Mariner 4 revealed Mars's scarred but canal-free surface in 1965.

the more volatile members of the public braced themselves for little green guys trying to say "Cheese." Others anticipated seeing some gnarly vegetation strewn across the ground. But even sober scientists were stunned by these first views of a barren plain, breaded with small rocks—a lifeless landscape apparently unchanged for billions of years.

It was a daunting sight, but consolation was quickly found in the fact that Mars's desolate looks could be deceiving. After all, not all biological activity is obvious in a landscape photo.

Anticipating life that might be hard to see, the Viking landers had been designed to be compact biology labs, capable of sophisticated chemical searches for creatures that might be microscopic or hiding beneath the surface. One of these tests—the so-called labeled release experiment—fed a liquid nutrient to a Martian soil sample, took a time-out, and then looked for exhaled gases. The nutrient, called chicken soup by the scientists, contained such tasty ingredients as formate, glycine, alanine, lactate, and glycolate (you will find rather few of these explicitly listed on a can of conventional chicken soup). But these compounds were built with radioactive carbon, an element that would be easy to recognize if the minuscule Martians had actually consumed the consommé and let loose with some gaseous waste products.

The experiments were run, the data were analyzed, and the Viking Biology Team weighed in with their verdict. It was thumbs down. Mars was no more hospitable to life than the moon; its surface was sterile. The culprits were extreme dryness and the relentless, deadly ultraviolet light of the Sun, a form of radiation that is reduced to mere sunburn strength on Earth by our thick atmosphere. Exhibit A in the verdict was a straightforward chemical analysis of the soil using a mass spectrometer, a well-known laboratory device able to measure the elemental and molecular makeup of any material. The onboard spectrometer failed to turn up organic molecules. Since organics are part of all life we know, the spectrometer's result was given great weight in concluding that the Martian landscape was deader than a pipe wrench.

One member of the biology team, public-health engineer Gilbert Levin, disagreed, and still does. Levin, the principal investigator for the labeled release experiment, claims that it produced results that were actually consistent with life, albeit life that might be somewhat different from the microbes that berth in the surface dirt of Earth. Indeed, a new analysis of the Viking mass spectrometer suggests it might easily have missed the Martians' molecular building blocks. It would have been unable to find the organic material in, for instance, the salt-laced soil of Chile's Atacama Desert, which is chockablock with microbes. This new work—while not proving that life was missed—at least reopens the door to that possibility.

Further space probes have shown that it's only one door among many. While the landers were pawing the Martian dirt, the two spacecraft that had brought them to the red planet were undertaking a high-resolution surveillance from orbit. These photos were, in their own way, as startling as the images from Mariner 4. They showed sinuous ravines everywhere, meandering channels that were reminiscent of river canyons in the American Southwest. Although they were dry now, at some time in Mars's dim past, liquid water must have gushed across its sands. If this was merely a sporadic event—an occasional flood lasting weeks or maybe a few years—then that would only interest geologists, not biologists. If, on the other hand, Mars once had lakes or oceans, they would provide long-term environments in which life could arise and flourish.

While the existence of ancient Martian oceans is still uncertain, probes have found growing evidence for primordial lakes. A camera on NASA's Mars Reconnaissance Orbiter has found geological clues to a onetime body of water the size of Lake Huron in what's now known as Holden crater. This watery feature was relatively short-lived—maybe a few thousand years. But could other wet spots have lasted longer?

In early 2004, two NASA rovers, Spirit and Opportunity, began rolling around the red planet. They came across multiple lines of evidence that standing water once pooled and puddled on Mars—water that could have jump-started life. In 2008, the Phoenix lander clawed up ice that lay mere inches below the Martian surface. Another NASA effort, the Odyssey orbiter, has espied salt deposits on the surface of the red planet, suggesting that billions of years ago the planet had periods when it was both far warmer and wetter than it is today.

The extent and persistence of liquid water that once drenched the landscapes of Mars are still topics that are being investigated. But of possibly greater import are clues that liquid water can be found in abundance on Mars today—not on the surface, but in dark aquifers about 300 feet (100 meters) or so beneath its ruddy epidermis. These underground reserves of moisture could provide a retreat for life now—an environment where small organisms could be protected from lethal desiccation and stinging ultraviolet attack.

As a result, the idea of life on the red planet has made a comeback. Thirty years ago, the Viking scientists were disappointed that Mars seemed as sterile as a mule. Today, both the lander experiments and the doleful conclusions drawn from them have been refiled in the "questionable" category. If you ask researchers today about the chance of finding life on Mars, you'll hear some restrained optimism. Martians might be uncovered if we dig deep enough underground.

If you ask those same scientists to speculate on what Martian life might look like, they'll suggest you grab a microscope, because you'll need optical assistance to see it. In the last half century, we've downgraded our nearest neighbors from the telepathic beings of Ray Bradbury's *Martian*

Chronicles to senseless, single-celled pond scum. Eking out an existence on Mars is clearly a tough slog for *any* type of life, and impossible for the sort of complex, intelligent beings that might construct canals or radio transmitters. Clearly, for those raised on sci-fi tales of elaborate Barsoomian societies as described by Edgar Rice Burroughs, this is a first-rate disappointment. But it doesn't mean that science has given up on the existence of sentient aliens. It only means that they aren't going to be found on Mars. Or, for that matter, on any of the other worlds of the solar system.

Mercury is airless and without water. Venus is hot and dry. The outer planets are mostly giant globes of gas, bundled up in heavy methane and ammonia atmospheres. Pluto is, as far as anyone knows, just a bantam ice ball—not even of sufficient majesty to be called a planet. And that's it. Sure, in the last years of the century, robotic spacecraft did show that several moons of Jupiter and Saturn might be capable of supporting life. This is one of the space program's most important discoveries. Nonetheless, the chances that any of these possible life-forms would be clever enough to converse with Earthlings (let alone be inclined to do so) is as good as nil.

It may sound rudely self-congratulatory, but scientific opinion is that *Homo sapiens* is the smartest species in the solar system.

RADIO WAVES FROM SPACE

In the domain of the Sun, we're still top canine. But with a few hundred billion other stars bobbing around the mammoth stellar carousel we call the Milky Way galaxy, there's plenty of unknown real estate—some of which might house the type of extraterrestrials that routinely provoke Mr. Spock.

Beginning a half century ago, anyone seriously interested in aliens adjusted their sights. They took aim at stars beyond the Sun. These other stars, and their putative planets, are, of course, remote. We measure the distance to Mars or Jupiter in tens of millions of miles. Those are big numbers, comparable to what a commercial airliner will typically take a lifetime to rack up. But distances to the stars are measured in light-years—one of which is about six trillion miles (ten trillion kilometers). (Working astronomers and sci-fi movie characters in search of some pseudoauthenticity eschew the term "light-years" in favor of another measure of distance, "parsecs." One parsec is about 3.26 light-years.) Even the closest stars are roughly a half dozen light-years away—many tens of trillions of miles. We cannot send robot probes to these neighboring stellar realms, checking out planets and hunting for aliens. It's too far for us now.

All of this was as obvious as Cyrano de Bergerac's schnozzle to scientists of the mid-20th century. So they considered ways to detect aliens other than simply blasting off and visiting their worlds. In particular, an idea floated earlier in the century soon became very appealing both in theory and practice: Listen for radio signals.

By the late 1950s, radio technology had advanced enormously beyond two decades earlier, when David Todd encouraged folks to tune in Martians in the comfort of their own homes. In particular, there was now apparatus that could operate in the so-called *microwave* region of the spectrum. Microwaves are high-frequency radio signals, with wavelengths that are typically a thousand times shorter than those of commercial radio broadcasts (hence the name). Microwave engineering had bloomed because of the development of radar.

Radar's emergence was prompted by the need to detect enemy planes during the Second World War. The use of radio waves as probes to find incoming aircraft was, in practical terms, similar to the wartime practice of lighting up bombers at night with powerful, Hollywood-style searchlights. This simple ground-based scheme was obviously limited to short distances, and it worked only in good weather.

Radar, which bounced radio rather than light beams off planes, wasn't subject to such limitations. In addition, by measuring how long it took for the transmitted beam to make the round-trip to the target and back, the distance to the aircraft under surveillance could be reckoned. With radar, you would get both direction and range. You could pinpoint the target, not just detect it.

There was a hitch, of course. For radar to work, technicians had to build equipment that could produce a narrowly focused transmitter beam. With a sloppy, big beam, you might still pick up a radio "echo" from some reflecting object, but you wouldn't know the target's precise direction. Plus, shorter wavelengths are easier to focus—or rather, they can be focused with antennas that aren't the size of a small ranch (which is more or less what you'd need if you used the longer wavelengths typical of AM radio). The high-frequency signals required for radar are the radio counterpart to the high-pitched squeals and clicks emitted by bats and dolphins for echolocation.

These ideas were obvious in the early 1940s, spurring the rapid development of microwave technology by all the major combatants.*

**The detailed history of radar—an acronym that was coined by the American military in 1941 from radio detecting and ranging—is a fascinating and inspiring tale, even if not yet the subject of a TV miniseries.*

Today, radar is used for commercial aviation, for nabbing freeway speeders and for mapping thunderstorms. Microwave oscillators (transmitters, essentially) are mounted in attractive sheet metal boxes and bolted onto the kitchen cabinetry of new homes, where they're known as microwave ovens and used to resurrect leftover pot roast.

However, another, less well-known microwave development also sprang from wartime activity. Radar antennas, with their transmitters *switched off,* were turned toward the heavens. The idea was not to beam a signal into space but to passively listen for any signals coming in the reverse direction. This newfangled effort to learn about the universe based on its radio properties developed into the research discipline known as radio astronomy.

Radio astronomy took astronomers by surprise. Even by the 1930s, a decade after radio sets had become a common item in living rooms, scientists hadn't given much thought to the possibility that the cosmos might emanate its own chatter. The first clear signals from space were picked up accidentally by Karl Jansky, an engineer for Bell Telephone Laboratories. In 1931, Jansky was tasked by his bosses with assessing the effect of thunderstorms on transatlantic communications. He detected plenty of static boiling up from bad weather in the tropics, but was disconcerted by an additional steady background hiss that came from much farther away—from the gas clouds of the Milky Way. For astronomers, the idea that our galaxy was somehow "on the air" was a thoroughly radical one. The *New York Times* put Jansky's discovery on its front page in 1933, but the prestigious *Astrophysical Journal* didn't run an article about cosmic static until 1940 (and the article wasn't by Jansky but by an acolyte radio amateur from Illinois, Grote Reber).

The early dismissive attitude toward radio astronomy can be chalked up to college physics. Straightforward computation shows that the amount of radio noise that a star produces is modest—certainly not enough to account for Jansky's surprising find. Eventually, investigators realized that while ordinary stars whisper only a bit of radio signal into space, hot gas can bray and shout. The hot gas between the stars was generating the hiss picked up by Jansky's antenna.

Over the course of the next several decades, other natural radio emitters were uncovered—among them, various sorts of dead stars. For example, when the largest members of the stellar bestiary run out of nuclear fuel, they collapse upon themselves to form black holes. These bizarre cosmic trash-mashers sport intense magnetic fields and stupendous gravitational forces—a combination of factors that cause them to whip nearby charged particles into a frenzy. The frenetic motion of these particles results in copious quantities of radio static, a phenomenon that, in the case of extremely massive black holes, produces an emitting source known as a quasar.

Pulsars, which are highly regular radio transmitters, are also grave markers for burned-out stars. These stellar corpses, which in life are only somewhat larger than the Sun, end their luminous existence by imploding to a small, hot ball the size of downtown Philadelphia. Once again, their strong magnetic fields stir up a radio cacophony for anyone who wants to listen.

Jansky and Reber had opened up a new view of the universe with radio astronomy—and revealed a sky that looks entirely different from the endless spray of stars we see with our optical telescopes. Within three decades of Jansky's first discovery, this novel work had unveiled objects, such as the aforementioned quasars and pulsars, that had been beyond

The 210-foot Parkes radio telescope in Australia is the Southern Hemisphere's largest antenna.

the conjecture and imagination of any astronomer. And because these listening efforts involved the same radio techniques that were so important during the war, many of the early discoveries were racked up by radar engineers.

As the 20th century passed its midpoint, another variation of this radio scrutiny of the skies was proposed: SETI. As with most innovative ideas, it occurred to several people independently, and nearly simultaneously. A physicist at Cornell University, Philip Morrison, was among the first, to not just think of the idea, but also to publish it.

LISTENING FOR ALIENS

In 1959, Morrison was toying with possible experiments using gamma rays, the most energetic form of light. He believed that measuring these rays might allow us to gain insight into some of nature's most violent events (exploding stars, for example). Sometime during Morrison's ruminations, another physicist, Giuseppe Cocconi, chanced to walk into the office and make an odd suggestion: Wouldn't these gamma rays be the perfect medium for interstellar communication? This "unlikely question," as Morrison later called it, was intriguing. However, the two academics decided to first consider whether other regions of the electromagnetic spectrum, such as ordinary (visible) light, or even radio, might better serve as couriers of alien messages.

They didn't ponder the problem long. Thanks to the work of the early radio astronomers, Morrison and Cocconi knew that space was kind to radio, at least the short-wavelength microwave variety. While interstellar gas and dust block or distort visible light, microwaves are given a free pass. They suffer relatively little from their traverse through the thin clouds of material that billow and float between the stars. Consider the Milky Way on a clear summer night. Its pale sweep and delicate complexity are imposing, but the view is deceptive. Thanks to the interference of small grains of interstellar matter, known to professionals as "dust," you can't see the crowded downtown neighborhoods of our home galaxy. Indeed, your view of the Milky Way is a censored view, screened by massive, dusty curtains. Radio astronomers were quick to take advantage of the fact that radio waves can slice through this interstellar sludge, and were soon using antennas, rather than mirrored telescopes, to chart the Milky Way's shape and size.

Microwaves not only zip through galactic dust unmolested, they also barrel through Earth's atmosphere as if it weren't there. That's a major practical advantage not only for radio astronomers, but also for anyone hunting for deliberate messages from space. Researchers can keep their observatories on the ground, where access is simpler and the food is better. For Morrison and Cocconi, all these facts led to a promising bottom line: Radio waves can safely travel from the surface of an alien planet to the surface of Earth, a distance of hundreds or thousands of light-years, without being hijacked by atmospheres or interstellar matter. Radio seemed the kind of technology that ET would use to get in touch, an idea that wasn't, after all, particularly novel. But for the first time, it was based on real science.

Morrison and Cocconi knew another important technical tidbit. Wartime astronomers had figured out that hydrogen was a natural radio transmitter. Now that's important, because hydrogen—the simplest, and therefore lightest, of the elements—is the major component not just of stars but also the material *between* the stars. Indeed, hydrogen is the principal constituent of *all* the tangible stuff in the universe, accounting for three-fourths (by weight) of normal cosmic matter. And while the hydrogen drifting in the cold, dark spaces between the stars may be lonely (although, as noted, it has some dust for company), it's not inert. Each hydrogen atom occasionally* emits a tiny amount of radio radiation at the microwave wavelength of 21 centimeters. These now-and-then radio waves burped by a few floating hydrogen atoms are weaker than a flea's knees, of course. But the enormous number of atoms between the stars ensures that the cumulative radio signal is prodigious: the triumph of the proletariat.

Indeed, the 21-centimeter hydrogen line is so astronomically useful that scientists believe that every advanced society—

**The emission is very occasional indeed. Any given atom will emit a bit of radio signal roughly once every 12 million years.*

no matter what world they're on—will have the hydrogen frequency engraved on the tuning dials of their radio telescopes. That hypothesis has an obvious, and felicitous, consequence. Perhaps you recall the hailing frequency on *Star Trek?* It was a special spot on the dial that all galactic races would use and monitor. As a Trekkie, I always wondered how lumpy-headed aliens drifting into Federation territory for the first time would know about this hailing frequency. After all, they hadn't been hailed before. But the neutral hydrogen line supplies the answer. It's nature's hailing frequency, the emergency test tone of the cosmos. It's as good as having billboards scattered through space alerting starships: "Use 21 cm on your transponder, good buddy. We'll be listening." Cocconi and Morrison knew this in 1959, and they presumed that all worthy extraterrestrials knew it as well.

The two physicists fleshed out the essence of their proposal. Radio signals could, in principle, make it from star to star, and they had guessed where to tune the transmitters and receivers. But was their idea practical? Radio waves are, after all, forms of light. They travel at light speed, but they also fade rapidly with distance. Sending radio signals from one star system to another might sound plausible around the water cooler, but could you actually do it? Wouldn't the required transmitter power be so enormous that even aliens would balk at the bill?

The question still worries a lot of people, judging by my e-mail. How could any signal from light-years away be detectable? Strangely, these same skeptical correspondents don't seem to find it remarkable that human eyeballs—only about an inch in size, and sporting a pipsqueak lens and no ability to make time exposures—can see light from stars hundreds and thousands of light-years away. Couldn't our largest antennas perform a similar feat for radio waves deliberately beamed toward Earth? Morrison and Cocconi did the math. They imagined aiming our most powerful military radar at our most sensitive microwave receiver. Then, in their minds, they separated the two until the transmitter was just barely detectable, thus establishing the maximum communication distance. It turned out to be dozens of light-years.

That's an astounding result, especially when you consider that *Homo sapiens* is still a radio neophyte. Morrison and Cocconi reckoned that, scarcely a generation after Marconi, we already had the requisite technology for sending, not broadcasts to Mars, but *interstellar* messages. If we could do so, wouldn't other galactic societies, more advanced than ours, already be doing it? Radio messages might be flooding the galaxy, subtle drumbeats from alien beings, passing uselessly through our bodies and awaiting notice.

In the fall of 1959, only six months after Giuseppe Cocconi had entered Morrison's office and asked his "unlikely question," the two academics published their seminal paper on SETI in *Nature*. They suggested that radio astronomers should look for artificial signals produced by alien transmitters on worlds around nearby stars. They also recommended 21 centimeters as the best wavelength at which to listen. They closed their paper by poetically addressing the skeptics: *"The probability of success is difficult to estimate,"* they conceded, *"but if we never search, the chance of success is zero."*

This if-we-build-it-they-might-come epigram eventually become a mantra for the SETI community.

Quick to heed their own advice, the physicists wrote to Sir Bernard Lovell. He was the director of what is now known as the Jodrell Bank Observatory, just outside Manchester, England, yet another wartime radar expert who had turned to radio astronomy. Two years earlier, his lab had completed construction of a giant radio antenna, the Mark I telescope—a 250-foot monster metal ear whose mechanism boasted parts from cannibalized wartime gun turrets. Cocconi and Morrison urged Lovell to point his antenna at nearby stars in a hunt for extraterrestrial broadcasts. But Lovell was uninterested and unamused, dismissing the idea as silly (a skeptical stance that, years later, he reversed).

SETI DEBUTS

The Cornell theoreticians weren't getting any satisfaction from the British, but unbeknownst to them, a young American astronomer, Frank Drake, would serve up satisfaction in spades just a few months later. Drake, who was barely 30 years old, had taken a postdoctoral position at the new National Radio Astronomy Observatory (NRAO) in Green Bank, West Virginia.

The observatory was in the process of designing and building a 140-foot radio telescope, an intricate process that would take years. The 1960s in general was the decade in which large antennas for radio astronomy would sprout across the country and beyond. The largest of these was the 1,000-foot-diameter Arecibo telescope, erected in a natural limestone bowl in northwest Puerto Rico. But that behemoth (which even today is the largest single-dish radio antenna in the world) was still three years from completion.

While the 140-foot telescope was being constructed, NRAO decided to forestall boredom and frustration among its astronomers by erecting a more modest 85-foot off-the-shelf antenna—a training-wheels instrument. The observatory director, Lloyd Berkner, a physicist and (yes) radar expert, pressed his staff to devise interesting projects that would exercise this relatively small dish. Drake, who had long been interested in the possibility of extraterrestrial beings, mulled over the challenge from his boss. He wondered whether the new antenna had any chance of detecting deliberate cosmic transmissions from alien worlds, unaware that this very idea was being written up by Cocconi and Morrison a few states away. Drake made the same calculations they had, and concluded that signals from stars within about a dozen light-years would be detectable. He also settled on the 21-centimeter neutral hydrogen line as the hailing channel, although mostly for practical reasons. He knew that any receiver operating at this wavelength would be doubly useful for traditional radio astronomy, and therefore none of his colleagues would grouse about the few thousand dollars necessary to build it. Drake and others began assembling the hardware for a scan of some very nearby stars.

Then, unexpectedly, the world shifted. The observatory acquired a new director, Otto Struve, a seasoned astronomer who was sympathetic to the idea that the universe might be crowded with cosmic company.* Immediately following his arrival, the article by Cocconi and Morrison hit the library racks. It was a first strike, the initial claim for the SETI intellectual mother lode. Struve knew that the NRAO needed to move quickly if it wanted to maintain any ownership of the SETI idea. He was also canny enough to realize that Cocconi and Morrison's theoretical work actually offered an opportunity. As he saw it, Drake's group could be the first not merely

---

**Struve was the editor of the* Astrophysical Journal *in 1940 who decided to publish Grote Reber's measures of cosmic static from the Milky Way.*

to hypothesize alien microwave communications but actually to launch a search. (He was unaware that Bernard Lovell had been asked and had demurred.) Hunting for signals was the difference between spending time in Spanish bars, thinking about crossing the Atlantic, and actually building the *Nina,* the *Pinta,* and the *Santa Maria.*

Struve began publicizing the observatory's plans to eavesdrop on possible alien broadcasts. A talk he gave in the fall of 1959 at an MIT lecture series drew the attention of John Lear, a writer for *Saturday Review.* Within weeks, Lear's article exposed Drake's plans to the public. This essay transformed a small, back-burner research program into a highly visible bid to finally locate some extraterrestrials. Adding a patina of substance and celebrity to the upcoming effort, Drake gave his experiment a name: Project Ozma.

"This was a search for exotic beings on a world far away," Drake explained. "That description matches the Land of Oz, the subject of some of my favorite books from childhood. So it seemed appropriate to name the project after the fictional princess of Oz."

In early April 1960, the observations began. Each day, in the frigid hours before dawn, Drake climbed to the focal point of the 85-foot and adjusted the electronics. The antenna was first pivoted toward the Sunlike star Tau Ceti—a dim object in the southern sky about a dozen light-years away. If Tau Ceti had planets, and if any of those planets were home to beings who were using powerful radio transmitters at 21 centimeters, Drake's eavesdropping apparatus might pick up a signal. For hours his team sat watching the pen of their chart recorder as it wobbled in response to the antenna's output. A small motor tirelessly twisted the tuning knob of the receiver up and down the dial, once a minute, but always staying close to the hydrogen frequency. As they huddled in a small, equipment-packed room, itself a speck lost between the softly rolling mountains that boxed in West Virginia's Deer Creek Valley, these pioneering alien hunters were expectant. History of a special kind could be made. All that need occur was a quiver of the recording pen. They waited for a sudden flounce, a quick jog betraying a signal—and, by implication, someone or something out there.

It didn't happen. Tau Ceti dipped below the horizon, mute to the last.

As planned, the antenna was swung toward the other star in Project Ozma's short list: Epsilon Eridani. This was also a Southern Hemisphere object, a star slightly dimmer than our Sun, and a few light-years closer to us than Tau Ceti. To facilitate the monitoring, the experimenters connected an audio amplifier and loudspeaker to their rig. Now they could hear the faint, noisy static produced by the telescope receiver. The minutes passed dully.

Suddenly, a loud, hammering signal erupted. Eight times a second, a whoosh burst from the loudspeaker, and the recorder pen masochistically banged against its mechanical stop. The unexpected had happened. Drake's first reaction was "What do we do now?" mixed with "Could it be this easy?"

The excitement was short-lived. The hammering abated. Less than two weeks later, it returned, but this time the astronomers had rigged up a second antenna, one aimed not at Epsilon Eridani but at the whole sky (including whatever aircraft might be flying over that part of West Virginia). The signal was also strongly picked up by this second, nondirectional antenna, and that meant that it wasn't coming from the position of the star. It waxed and waned in strength, just as would be expected from an aircraft transmitter.

Some of the media had already heard of Drake's detection. He explained that despite the excitement, he had only found intelligence on Earth, not on a planet ten light-years away. It was the first of modern SETI's many false alarms.

Within two months, Project Ozma was over. It hadn't uncovered the aliens, but it had shown how one might do so. It was, in every sense, a truly pioneering effort—incorporating, as it did, almost all of the basic ideas and technical choices that would be used by its followers.

One of those followers was Project Phoenix, which was begun 35 years after Drake's modest two-month effort. Phoenix was the most ambitious and sensitive SETI search ever made, and it was undertaken by a new organization—the SETI Institute—founded on November 20, 1984. Rather than examining two Sunlike stars, as Ozma had done, Phoenix had a planned target list of a thousand. Instead of monitoring one frequency at a time, scanning up and down the dial with a motorized knob twister, this new effort listened to tens of millions of channels simultaneously. And finally, in place of using an 85-foot antenna to funnel signals to its sensitive electronic amplifiers, Project Phoenix lined up the world's biggest antennas: the 210-foot Parkes radio telescope in Australia; the 1,000-foot Arecibo dish in Puerto Rico; and the completed 140-foot scope in Green Bank—located only a few hundred yards from the instrument that Frank Drake had used to usher in a new way of finding beings beyond Earth.

It was while using the 140-foot Green Bank antenna that we found the interesting signal described earlier. And while that 1997 signal turned out to be a false alarm—a signal from *Homo sapiens* rather than from the inhabitants of another world—the landscape for SETI was changing rapidly.

WHAT WOULD BE THE MILITARY RESPONSE?

Many people are confident that, sequestered deep within the cryptic machinery of the United States Department of Defense, are plans and personnel ready to deal with the aftermath of discovering extraterrestrials. It's a pleasant assumption, and one that gains currency from the fact that, a half century ago, both the U.S. military and the Central Intelligence Agency got their collective knickers in a knot about UFOs. Both figured that there was at least a chance that some UFOs were uninvited visitors to our world. However, they worried more about uninvited Soviet aircraft or missiles. The government did, indeed, try to formulate plans.

So it's true that the military has shown interest in extraterrestrials, in particular, those that might be violating our airspace. But detection of a signal from another planet, around another star, would be a totally different kettle of kippers. To begin with, SETI is a passive listening experiment. The aliens wouldn't know that we'd picked up their broadcasts any more than the BBC knows that you've just tuned in one of their short-wave news reports. Doing so hardly ever provokes BBC announcers to leave their London studios, seek you out, and either destroy your neighborhood or abduct you for experiments that are too personal by half. So it's hard to imagine that the military would see any security threat in a detection, a result that's akin to finding a bottled message washed up on the beach.

There's also the barrier of distance. While we don't know how far the nearest aliens are, it seems unlikely on statistical grounds that they'd be any closer than a few hundred light-years. That's an enormous distance, even for the most advanced rockets we can imagine. The aliens, whether we detect them or not, are remote, and we are buffered from their predations (if that's what they have on their minds) in the same way that oceans buffered Native Americans from being bothered by the Roman Empire.

Of course, you could rightly argue that the first residents of the Americas eventually were bothered, once the Europeans developed ship technology that could cross those insulating seas. Maybe the aliens have the equivalent of warp drive, and can use exotic physics to bridge vast interstellar tracts. Maybe our early television and radar signals have given aggressive aliens a head start and they're already on their way to trash our world or haul you out of your bedroom. In either case, shouldn't the military be prepared for actual visitors?

That would be a waste of your taxes. It's unclear whether humans will ever be able to send themselves to the stars—the technical barriers are formidable. But even optimists would concede that such abilities are at least a few hundred years in our future. Consequently, any extraterrestrials who come here can be safely reckoned to be centuries or more ahead of us, technologically speaking. Any earthly defense against such a society would be like the armies of the American Civil War making plans to battle the U.S. military of today.

Hollywood occasionally tries to circumvent this dead-obvious circumstance by invoking human bravery or cleverness for a Hail Mary defense against invaders from space. But bravery doesn't help much if your weaponry is outclassed by a few orders of magnitude. And outwitting the extraterrestrials doesn't make sense. In the popular 1996 film *Independence Day,* our species turns the tables on some nasty aliens by uploading a virus to their computers. This is absurd to the point of being quaint. Imagine a "computer" of a century or more ago, for example, Charles Babbage's difference engine, being used to disable a modern laptop. The two machines are of completely incompatible construction. Indeed, about the only way such an early machine could cripple your laptop is if the former was dropped on the latter.

One can argue that the military has devised plenty of contingency plans of which we're unaware, and a defense against aliens might be one of them. But assuming this to be true is merely paranoia fed by irrational argument.

WHY WOULD ANYONE BELIEVE THEY'RE OUT THERE?

Jill Tarter and I were once invited to talk to talented high school kids in Winston-Salem, North Carolina. I had never given a lot of thought to North Carolina. For most of my life, it had merely been a 150-mile (240-kilometer) barrier of pine-covered seaboard, blocking the way to Florida. Winston-Salem is a tobacco city and, as the locals repeatedly point out, the home of Krispy Kreme Doughnuts, a product perceived, although not proved, to be less lethal. We came to discuss our day job: the search for intelligent life in the cosmos.

The kids were enthusiastic, as kids usually are given this subject. Youngsters love aliens in the same way they love dinosaurs: Both are exotic, scaly, and no threat to them whatsoever. The dinosaurs are dead and the aliens are distant. Having spoken to the students, we were to wind up our stay in Winston-Salem with an evening lecture for the general public.

Jill's first slide lit up the large center-stage screen as she described how the universe was chockablock with the ingredients for biology, and possibly with life itself. My own presentation, which followed, elucidated our attempts to find the more intelligent varieties of that possible life. The audience was attentive.

For the question-and-answer period, two lecterns, one on each side of the screen, were dragged in. Jill and I, standing behind our respective rostrums, took turns with the questions. Mostly, they were the usual interrogatories—technical queries about the experiment, puzzlement as to why we weren't broadcasting, and our opinion about what happened at Roswell, New Mexico (about which, more later). Then a young man stood up near the back and was passed the wireless microphone. Since it was my turn, his question was directed to me.

"You're talking about possible beings on other planets," he began, with a hint of aggression in his voice. "But they cannot exist. Scripture tells us that."

I looked to my left, to Jill's lectern. She returned my glance, smiled, and sat down.

What followed was a low-key harangue from the youth, peppered with biblical chapter and verse, as he challenged me to refute an obvious truth: We were on a fool's errand, looking for extraterrestrials. They could not exist, knowledge that comes to us from the highest authority. Not only were we wasting our time and somebody's money, but we were also guilty of hubris, of challenging God. This went on for five or ten minutes, during which time I silently gripped the lectern and the audience made discomfited noises.

You see, there's more to Winston-Salem than coffin nails and crullers. It's part of the Bible Belt.

The comment was tough to handle, and for many of the same reasons that questions about UFOs are difficult. The questioner barraged me with specific passages, straightaway crippling my response. I didn't know the quotes, and by not responding to them in detail I appeared ignorant—and therefore guilty as charged. If I wasn't familiar with the citations from the Good Book, how could I possibly disagree with their premise? Death by detail.

My response would have to be diplomatic and delicate. It was also brief. I simply said that the universe has routinely surprised us with its richness and deep complexity. Quasars, pulsars, and even giant clouds of galaxies are all unexpected novelties, hidden to both the personae of biblical times and the astronomers of a century ago. I saw no reason to reject wondrous phenomena simply because they are unheralded in the Bible. More to the point, it struck me as unconscionable not to use our intellect—truly a gift worthy of God—to explore the universe, and possibly find companionship. This was, of course, a paraphrase of what Galileo had written to the Grand Duchess Christina a few centuries earlier, to wit: "I do not feel obliged to believe that the same God who has endowed us with sense, reason, and intellect has intended us to forgo their use."

My response was greeted with diffident applause. After the event, many people, some motivated by embarrassment, made a point of hanging around to personally thank the speakers.

Frankly, audiences do not often object to the very premise of a talk on SETI or, more generally, life in space. The people who come to my lectures are usually inclined to think that the universe could be rife with life. So the negative reaction to SETI encountered in Winston-Salem was unusual, at least in my dealings with the public. But what's the reaction of academics? I'm frequently asked how mainstream astronomers feel about spending money to point telescopes at the sky, not as an effort to learn more about nature, but in a hunt for hairless gray guys with big eyes.

The Cambridge pulsar folk resorted to a chuckle-inducing description of putative aliens as "little green men." That might incline you to believe that astronomers regard SETI experiments as a waste of time. Some astronomers probably do think that way, particularly if they believe that the SETI crowd is competing with them for grant monies or access to the telescopes. After all, if the astronomers get their hands on those instruments for research, they're virtually assured of coming back with data—a photograph or a spectrum—that can be examined for new insights. They gain fresh, publishable results, useful for impressing peers and winning lifetime employment at the local university. SETI scientists always come back from the telescopes as they went: with empty hands.

That contrast is indisputably true. SETI has booked only negative results so far. But in my experience, few astronomers have such a simplistic attitude. I've encountered no more than a handful who doubt that the extraterrestrials are out there. Nearly all concur with premise one of the SETI enterprise, namely that both life and intelligent life abundantly suffuse the universe. Not all agree it's worth the effort to search. Some dispute that our techniques are sensitive or comprehensive enough to have any chance of success. We might even be using the wrong physics. Others figure that intelligent life is very rare, and no matter what technology we deploy, we won't find it.

Nonetheless, the decadal reports on astronomy produced by the U.S. National Academy of Sciences—documents drawn up by astronomers for the purpose of setting research priorities—have endorsed SETI for 40 years. The 1972 statement on this enterprise is typical:

"More and more scientists feel that contact with other civilizations is no longer something beyond our dreams but a natural event in the history of mankind that will perhaps occur in the lifetime of many of us. . . . In the long run, this may be one of science's most important and most profound contributions to mankind and to our civilization."

The idea that extraterrestrial contact might be imminent and important sounds nice, but what motivated the National Academy to say so? It's not just a subconscious reaction to 200 episodes of *The X-Files* or a secret fondness for Kepler or Lowell. The report's conclusion is based on a concatenation of facts that every astronomer knows—facts determined during the past two centuries of astronomical research.

First fact: The physics and materials of the universe are the same throughout. The most distant stars in the most distant galaxies are composed of elements tabulated and enumerated on wall charts in every high school chemistry class. It's the same stuff that makes up the Sun, the Earth, and your pet ferret. You might wonder how we could possibly know this, given the prodigious distances of space. Starlight provides the evidence. For more than a century, scientists have studied the dark lines that blemish the rainbow banner produced when a prism—or its modern high-tech equivalent, a spectroscope—breaks up the light from a star. The dark markings arise when the star's light is selectively absorbed by gases that wrap its relatively cooler, outer regions. The lines are as characteristic as the hallmarks on Reed & Barton flatware.

In Victorian times, astronomers became adept at identifying the elements causing these varied absorptions—hydrogen, helium, nitrogen, calcium, sodium and much, much more—long before physicists had puzzled out the quantum physics responsible for the lines' detailed characteristics. That chronology was similar to the work of biologists, who laboriously classified plants and animals despite being ignorant of the function, let alone the structure, of DNA. But just as DNA was the key to grasping the true relationships among living things—providing a mechanism to elucidate a century's worth of taxonomy—so too was quantum physics the key to understanding the fingerprints of the stars. In the 20th century we unraveled those spectral codes. From them we learned that the stars in galaxies tens of billions of light-years away are constructed of the same elements you'll find in the Sun.

The simple truth is that the rules and raw materials of the universe are everywhere identical. Physics and chemistry are truly universal, which will spare you from having to take these classes a second time should you move to another galaxy. Consequently, what happened here—namely the appearance of life—could have happened anywhere.

The second relevant fact, and one that many regard as more persuasive than the Oxford debate team on a good night, is the enormous size of the cosmos. A typical galaxy, such as our own Milky Way, houses a few hundred billion stars. The latest estimate for the number of galaxies just in the portion of the universe we can see is about 80 billion. Multiply these two numbers on a pocket abacus, and you'll be confronted with the impressive fact that the tally of stars within our purview is about 10^{22}, which is a one followed by 22 zeroes. That number makes everyday ciphers, such as the national debt, seem inconsequential. Put another way, the sum total of all stars within the visible universe equals the number of glasses of water in all the oceans of Earth.

Many scientists are reluctant to believe that only our particular glass has spawned creatures able to write music, play mah-jongg, or fathom the way the universe is put together. That would make us stunningly special. Of course, there are people (viz: Winston-Salem audience members) who think we *are* special. It's ungentlemanly to say so, but they could be wrong.

WORLDS APLENTY

The third applicable fact that scientists cite when asked about ET's existence derives from one of the great triumphs of recent astronomical research: the discovery of planets orbiting other stars. Since the first was found in 1995, more than 300 extrasolar worlds have been uncovered, mostly in our galactic backyard. Planets seem to be as common as freckles in Ireland.

This circumstance is so well known that it's become just another boring science fact. But it's neither boring nor trivial. The discovery of planets orbiting around other stars is the stunning finale to a long run of controversial speculation. In 1644, René Descartes wrote in his *Principles of Philosophy* that the stars were also suns. Each of these was presumed to have its own retinue of planets and, perhaps, inhabitants with souls. At the time, this proposition was a clear step over the line. Sure, Copernicus had already attacked conventional wisdom a century earlier by giving Earth the pink slip as the alpha male of the cosmos. He demoted the Earth, and the center-of-the-universe job went to the Sun. But Descartes brazenly suggested that Sol wasn't important, either. It was merely one star among many—possibly infinitely many. Descartes's belittling of the Sun was eventually ratified by the astronomers. Our star was just another member of the stellar horde.

Descartes's extended hypothesis—that these other stars are also haloed by planets—was harder to verify. The reason is uncomplicated: Planets shine by reflected light. This secondhand sheen is enough to make nearby Venus, Jupiter, and several other planetary brethren conspicuous from your backyard, but their luminance at the distances of the stars would fall short of easy notice. If you were an alien astronomer looking at the Sun in hopes of finding orbiting worlds, the most visible quarry would be Jupiter. But even this oversize orb would be, at best, a billion times fainter than the Sun. Earth would be ten times dimmer yet. From Alpha Centauri, our nearest stellar neighbor, the task of espying Earth would be equivalent to noticing a mosquito circling a lightbulb 25 feet (7.6 meters) away, as seen from 10,000 miles (16,000 kilometers) off.

So even though many astronomers felt in their bones that Descartes was right, and that planets might be plentiful, it was only a feeling: They couldn't actually see any. When Frank Drake conducted his first SETI experiment, no one had definitively found a planet circling around another star. That was still true three decades later when I joined the SETI Institute. We all just assumed these worlds (and their alien residents) were out there.

Assumptions might be appealing to Titian, but they're not adequate for science—at least, not indefinitely. Astronomers of the late 20th century, armed with improved technologies, kept working on finding worlds that orbited distant suns. Simple calculations proved that rather than trying to see the analogues of mosquitoes around lightbulbs, the researchers would do better to hunt for the slight wobble a planet induces in its home star. After all, stars and planets engage in a gravitational dance, and both move around the floor. But unlike the planets, the incandescent stars are easily visible.

So the question became, how might one actually detect this subtle stellar shimmy? The most sensitive approach would be to look for slow-speed changes in the absorption patterns in a star's spectrum.

This scheme looked good on paper, but in practice boiled down to hardware: making a spectroscope that was stable enough to detect this delicate stellar dance. Remember, an astronomical spectroscope is a souped-up prism that breaks up starlight into a rainbow of colors. But superposed on those colors are thin dark lines, appearing at the wavelengths of light that are absorbed by gases in the outer regions of the stars themselves. If the star is moving toward or away from us, those thin lines shift slightly, due to what's called the Doppler effect. Simply stated, if the star is moving toward us, the lines move toward the blue end of the stellar rainbow. If the star is receding, they move toward the red. This is analogous to the changing pitch of a train whistle as the locomotive either rolls toward you or glides away.

The problem of building a better spectroscope was solved by a group of Canadian astronomers in the early 1980s. They bolted a small glass bottle containing poisonous hydrogen fluoride gas into the light path of their telescope. The gassy vapors would produce their own absorption lines in the spectra (if they didn't leak out and kill the astronomers), and these would serve as a rock-solid reference grid for the actual stellar lines. This increased precision would, they hoped, allow them to discover the very subtle stellar dance that would betray the presence of planets.

The Canadians used their gas bottle to look for wobbles in 16 stars. They didn't find any. Had they observed another few dozen, they might have gained the enduring satisfaction of knowing their names would be typeset in textbooks for hundreds of years. They would have been the first to clearly locate a planet around another star, beating the first actual detection by a decade.

As it happened, that latter discovery—and it made front-page news everywhere—occurred in 1995. Two Swiss astronomers, Michel Mayor and Didier Queloz, weren't really looking for planets, and that actually worked to their advantage. Other researchers who *were* hunting for planets implicitly assumed that these undiscovered worlds would share the traits of the planets in our solar system. These take anywhere from 88 days (Mercury) to 164 years (Neptune) to orbit the Sun. The planet the Swiss accidentally found hoofed around its home star, 51 Pegasi, in little more than four days! This wasn't a sedate gravitational waltz: It was a high-speed honor-your-partner do-si-do.

This new world was completely unlike any we knew. Six times closer to its sun than Mercury is to ours, and with a mass at least half that of Jupiter's, this planet, creatively called 51 Peg b, was not only the first world found around another, ordinary star, but the first of a whole new class of planets called hot Jupiters: big planets sitting close to the fire of their suns.*

**In the hoopla that accompanied this discovery, everyone forgot that, four years earlier, astronomer Alexander Wolszczan had used a radio telescope to uncover a wobble in the pulsar PSR B1257+12. This wobble was a telltale sign of planets. Since a pulsar is a pathological object—a highly collapsed dead star—this first extrasolar planet detection was not accorded the same fame as the 51 Peg find.*

Nine months after this discovery, at a meeting of astronomers and biologists on the isle of Capri, the attendees were still buzzing over the news. Worlds around other stars were no longer just a plausible assumption, there were real data. I had lunch with Michel Mayor during the Capri meeting, and asked him—somewhat whimsically—why Switzerland even had observatories. I had realized at a relatively tender age that the countries that traditionally excel in astronomical research—the United States, the Netherlands, the United Kingdom, France, Italy, Germany, Russia, and a few others—were almost all major seafaring nations. That was hardly coincidence; astronomy once had important applications for navigation, especially the ability to establish your position on the oceans' vast, featureless tracts. Centuries ago, determining longitude required accurate timekeeping and a good astronomical ephemeris. If you had observatories, it was probably because you also had a navy.

"So," I inquired of Mayor, "why does Switzerland do astronomy? Was there a problem finding the other side of Lake Geneva?"

Mayor smiled. "No, it wasn't for navigation. It was because of the watch industry."

The watch industry?

Mayor continued: "Our country has a tradition of making expensive watches, and it was important to the manufacturers that these costly devices keep precise time. The watchmakers needed an accurate time standard. That meant astronomers."

I could imagine that after doing nothing but checking meridian transits each night to keep the clocks on track, those early Swiss stargazers became mammothly bored, and opted to do some research. Now their successors had found a planet around a Sunlike star, 50 light-years away. They had bedazzled the astronomical world.

Indeed, after 51 Peg, the dam burst. Other researchers began foraging for wobbles, most notably a team at the University of California, Berkeley—Geoff Marcy, Paul Butler, and Deborah Fischer. Within a handful of years, this group had discovered more planets than the total known to *Homo sapiens* since it first turned eyeballs on the sky. The planet orbiting 51 Peg was a type of world that's easy to find, albeit hard to predict. But once researchers knew about hot Jupiters, throngs of similar, steamy leviathans began rolling out of the spectroscopes. The wobble method, after all, is most sensitive for (and therefore highly biased toward) quickly uncovering large planets, close-in to their star.

In the tally so far, roughly one in five of all the discovered exoplanets (the name for planets outside our solar system)

An artist's conception of a "hot Jupiter" around another star

is a hot Jupiter. Most theoreticians figure that these massive worlds are actually born far from their star—where there's more gas and dust available for their construction. Then, due to various frictional effects, they spiral inward to end up as bloated, hot gas balls, stalled in small orbits around their sun. Typically, these simmering planets are hotter than Venus, and just about as livable.

As the 20th century rumbled to a close, astronomers began to uncover worlds that differed from these jumbo star-huggers. One reason for the increasing diversity was that the instruments were improving. Another was simply that the planet hunters had been looking for so long. With extended monitoring, slower wobbles eventually became noticeable. As an example, in our solar system Saturn takes 29 years to orbit the Sun. That's not the type of wobble you're likely to spot in only a few months at the telescope; the dance is too slow.

So those folks who actually thumbed the science sections of their newspapers noticed that the planetary bestiary was being enlarged with hot Neptunes (smaller versions of hot Jupiters) and super Earths (planets between one and ten times the mass of Earth). Today, when astronomers drive back from mountain tops with freshly bagged planets metaphorically strapped to the roof, they're most commonly Jupiter look-alikes—that is, more or less the same size as Jupiter, but in orbits that are many times larger than those of their hot brethren.

The planet hunting business has become an industry, with no downturn in sight. Let's look at the record: Uranus was found in 1781. Another 65 years elapsed before Neptune was discovered, and 84 more before Pluto's debut (it enjoyed 76 years of planethood). But on average today, a new planet is reported every week or two. While this is mostly an activity for observationalists, the strange nature of many of these extrasolar systems has given theoreticians plenty of fodder for publication. More than a dozen years of planet-seeking activity has spawned a spate of books and a ponderous stack of magazine articles. It's been an astronomical revolution.

A WORLD FOR ET?

This promising start indicates that planets are out there—and in abundance. What sort of abundance? After all, when an astronomer observes any given star, the chance that it will show wobbles is slight. So I asked Geoff Marcy—who has snared more planets than any other humanoid—how that success rate would change if the telescopes were absolutely perfect. How many planets are *really* out there?

"Well, current technology can efficiently detect the largest exoplanets, ones that are similar to Saturn and Jupiter," Marcy said. "And we find that 15 percent of all stars have such giant planets. Meanwhile, 75 percent of young stars show disks of gas and dust—planets in the making—and these disks may eventually give birth to small, rocky worlds."

Studying the disks that surround young stars, those that are only a few million years old, has become fashionable of late. Some of the disks are seen to have gaps—places where the dust is missing. The gaps are presumably caused by new planets which, as they form, sweep up the dust like a vacuum cleaner.

Combining the disk results and the stellar wobble data, Marcy concludes, "It seems likely that 50 to 75 percent of all stars harbor planets, the majority being somewhat Earthlike."

For an astronomer, to whom factors of two are seldom important, this fraction is virtually the same as "all."

Marcy is saying that essentially all stars have planets, and since planets (like kittens) come in litters, it's fair to say that the number of cool worlds in the visible universe is even greater than the number of stars. Even *more* than the number of glasses it would take to hold all the water in Earth's seas.

So that's encouraging news for an alien hunter, because it's a general presumption that life will evolve on a planet or moon. This requirement is a consequence of the building blocks making up all earthly life: carbon-based molecules. These molecules are assembled into yet more complex structures: proteins, for example, or the DNA that serves as the blueprint for everything that crawls, floats, flies, or sprouts. At temperatures even modestly above what we consider comfortable, carbon-based molecules break apart. At the gelid, bitter temperatures of interstellar space, chemical reactions are slow, making difficult the sort of rapid metabolism associated with intelligent creatures. Only planets and large moons, which might be blessed with moderate temperatures, seem suitable hosts for the evolution of advanced life. Of all the ordinary matter in the universe, planets and moons tally less than one percent. But they're likely to be where the action is, when it comes to life.

Now, you might argue that these assumptions reflect a lack of imagination, that they're shamefully anthropocentric. Why are we so wedded to "life as we know it," or more precisely, "intelligence as we know it"? That's a decent question, but we have good reasons for the restraint. Sure, sophisticated life might arise in less familiar environments, for example, the thin, gassy expanses of an interstellar cloud or the brittle surface of a neutron star. These possibilities have been explored in science fiction by authors who knew their science. So clearly, not everyone thinks such oddball scenarios are impossible.

Nonetheless, the usual modus operandi of astrobiologists (scientists who busy themselves with the matter of extraterrestrial life) is to be conservative, to be unabashedly terra-centric. Nature is undoubtedly ingenious beyond our impoverished imaginings. But by assuming that complex life needs the type of environments found on Earth, we can be sure that our postulates have not exceeded the possible. We haven't fallen into the trap—so frequently set in this business—of admitting that we don't know everything, and therefore anything could happen. In addition, using the single example of intelligent life apparent to us provides guidance in what to search for, and where. You can imagine anything you want, but it's hard to design an experiment when all things are equally possible.

This brings us immediately to a problem. Virtually none of those 300-plus extrasolar planets discovered so far is like Earth. Or at least, none is the same *size* as Earth. They're all bigger, and usually tens or hundreds of times bigger. In comparison, Earth is a rocky runt.

So where can Earth's planetary brethren be found? The answer, one you'll get at any cocktail party blighted with astronomer guests, is simple. Just as asteroids and comets greatly outnumber planets, because they require less raw material, small planets are likely to outnumber large ones. The next question is: Do we have hard data to back up the optimism?

Astronomers have already detected those aforementioned "super Earths," planets bigger than ours, but small enough you could tolerate flying from one hemisphere to the other in a coach seat if you were a native. Super Earths, which range from the mass of our planet to one that's ten times heftier, are likely the only ones to have breathable atmospheres. Smaller worlds wouldn't have enough gravitation to retain their air, and bigger bodies would keep too much, producing the type of heavy, acrid atmospheres that swathe Jupiter and Saturn.

So far only a few dozen such worlds have been found. But several of these bulked-up planetary brothers are situated at that felicitous distance from their sun at which surface water would remain liquid—which means they might have oceans. By 2008, Didier Queloz, one of the two astronomers who found the planet around 51 Peg back in 1995, was claiming that at least one in three stars have super Earthlike worlds. The *really* good news (if astronomical bulletins can be so characterized) is that improved ground-based spectroscopes, such as the HARPS (High Accuracy Radial velocity Planetary Search) instrument that's mounted on an 11.8-foot telescope at the La Silla Observatory in Chile, are able to measure wobbles that are barely more than a slow-speed drift. According to Queloz, we will soon develop instruments that can measure a stellar shuffle amounting to no more than 0.4 inch (1 centimeter) per second, or slower than an ant. That's roughly 50 times better than today's capabilities. Such accuracy would easily allow us to find Earth-size (and smaller) worlds with our ground-based telescopes.

As this steady refinement of spectroscopes continues, the number of moderately sized worlds discovered by the planet hunters continues to increase. Indeed, while hot Jupiters and their ilk have dominated the headlines for a dozen years, Queloz and others are now proving what all nondelusional astronomers have always assumed: Most planets are small. It seems only a matter of time—the estimate is a few years—before the wobble-watchers find a planet the size of Earth.

NASA is taking another approach to finding these terrestrial analogues. Its Kepler mission, planned for launch early in 2009, features a spacecraft whose sole goal in life is to find worlds of modest girth. It will do this by doggedly staring for nearly four years at a single patch of sky. The patch, wedged between the bright stars Vega and Deneb in the constellation Cygnus, covers the same amount of sky as 400 full moons, packed limb to limb. Kepler's monotonous job is to keep track of the brightness of about 170,000 stars within this area.

If—as we expect—many of those stars have planets, then a small fraction, about 1 in 200, will be seen to experience mini-eclipses. The planets will occasionally pass in front of their suns, dimming the starlight ever so slightly.

The process works this way: Imagine staring at some streetlights a few blocks distant. Now consider what happens if a fly passes in front of one of those lights. You can't see the fly; the lamps are too far. But if you keep accurate tabs on the brightnesses of the lamps, you'll notice a very slight dimming because of the intervening bug. The amount of dimming will tell you how fat a fly you've found.

Assuming all goes well, within months after launch Kepler will uncover hundreds of hot Jupiters. Kepler could quickly treble the number of known extrasolar planets. But finding more jumbo worlds, while easy, is not the point of launching Kepler. During the years of its mission, NASA's space-based scope is expected to detect somewhere between a few dozen and a few hundred planets that are roughly the size of Earth.

Thanks to Kepler's efforts, we'll then know what percentage of stars have Earth-size worlds, and what fraction of these are in orbits that keep them warm enough for liquid oceans but not so hot as to turn their seas to steam. This is a space mission with a difference: It's rather akin to the generation of explorers that scoured the globe after Columbus's discovery of America. Within a half century or so, our world maps were largely filled in; the globe had ink all over it. No previous generation could have done that, and no following generation had to: Since then, all we've done is refine a few details. In a similar fashion, Kepler will establish what solar systems throughout the universe look like—once and, in some sense, for all.

Of course, it might turn out that all our optimism is counterfeit. Truly Earth-size planets could be rare; solar systems like our own might be exceptional. Few astronomers think the first is true, and there's no compelling evidence for the latter, either. But the proof can be ascertained only by observation. At least you won't wait long before the jury returns; the verdict will be in well before the 21st century has passed its teenage years.

In light of this work, it's sobering to recall that scarcely 20 generations ago, learned men firmly believed that the Earth was both singular and central. Our planet was privileged because it was different. Everything in the sky, other than the moon, was luminous: The Sun, stars, planets, and comets all glowed with light (or so it seemed). Earth at night, on the other hand, was dark. More than that, the appearance of these heavenly bodies was always the same, unlike Earth, where everything from weather to warfare took place. Clearly, our planet was unique, the center-stage attraction for the cosmos.

Today we may be on the verge of showing that our world, at least geologically, is about as prosaic as pigeons. Just another member of a teeming planet flock, of which many will surely be of similar composition and circumstance. It's speculative but hardly radical to suggest that the number of Earthlike worlds in our galaxy alone tallies in the tens of billions.

BUT IS THERE LIFE?

Recent results from astronomers don't lend much support to the notion posited in Winston-Salem, namely that ET couldn't exist. To the contrary, the stargazers have found reason to expect many theaters in which the drama of sentient existence could be playing. For many astronomers, that conclusion wraps up the discussion. From their point of view, charting the sky is the heavy lifting, and all the rest—the small matters of life's origin and the evolution of intelligence—is just inevitable, dirty chemistry.

This short-circuited thinking—that having a plethora of worlds like Earth essentially guarantees many instances of intelligence—often rankles biologists. Their field of study is far messier than that of the astronomers—less dictated by the ruthless, ever present laws of physics, and more susceptible to the vagaries of chance and contingency. Astronomy is law and order; biology is a raucous free-for-all. So not all life scientists believe that merely having gobs of agreeable planets ensures hordes of cosmic confreres. Biologists don't hesitate to point out that we've still not found *any* conclusive evidence of life elsewhere—even microscopic life—let alone the type that's clever enough to hold up its side of the conversation or abduct you for crossbreeding experiments.

Nonetheless, recent research into the prevalence of living things on Earth has nudged many biologists toward restrained optimism. Exploration has shown that life can be found in just about every locale within two miles (three kilometers) of sea level. Certain types of life, in particular the microscopic variety, are able to survive under conditions we'd classify as cruel and unusual. Ergo, just as a tolerance for a wide range of foodstuffs increases the likelihood that you'll frequent a broader range of restaurants, so too does the discovery of life's remarkable ability to adapt suggest that it could be present on worlds that, at first glance, might seem bluntly inhospitable.

Consider a currently fashionable category of life known as extremophiles. True to their name, extremophiles can batten and fatten in conditions that humans—and most other species—would consider off-limits. The first of these sturdy organisms, a thermophile, was found in the late 1960s in Yellowstone National Park, hanging out in one of the hot springs. This bacterium was given a name bigger than itself: *Thermus aquaticus*—literally, "warm bath water dweller." (Species names, by the way, are impressive only if you don't know Latin.)

Thermus aquaticus could not only endure but thrive in temperatures above 160°F (71°C). For comparison, spin your hot water tap and let it run. The water will scald your hand, but the temperature won't exceed 140°F (60°C). This is observational proof that you are not a thermophile.

Yet *Thermus aquaticus* is only moderately hardy when it comes to taking the heat. One hyperthermophile, *Pyrolobus fumarii,* can tread water at a scalding 235°F (113°C). That's not only above boiling point; it will soften the upholstery in your pickup. Other extremophiles operate in below-freezing cold (psychrophiles), highly acid or base solutions (acidophiles and alkaliphiles), heavy-duty brines (halophiles), and in circumstances of crushing high pressure or dusty dryness (piezophiles and xerophiles). Some varieties can shrug off nuclear radiation, or dwell happily in aviation fuel. Frankly, extremophiles would be recruited for the local SWAT team if they were big enough to carry weapons.

How these tiny tough guys survive in such abusive abodes is an entirely separate field of study. But the point is that life has proved itself more adaptable than a little black dress. Think about it for a moment: In the past three or four billion years, Earth has tolerated massive changes in its atmosphere, withstood pummeling by hulking rocks from space, suffered sudden blasts of radiation from both the Sun and from distant cosmic events, undergone reversals in its magnetic field, and endured repeated ice ages. The tree of life has been cut back, pruned, and generally abused. But in all that time, life hasn't gone away. Life is durable.

This is an important touchstone when trying to guess if intelligent beings are common or rare. If life were less plastic, less able to roll with the unavoidable environmental punches (not to mention the threats caused by its own destructive tendencies), then it might be routinely extinguished long before becoming sufficiently complex to spawn creatures with intelligence.

It seems reasonable to assume that this degree of plasticity is not anomalous, but the inevitable consequence of simple feedback. If a species has a mutation rate that's too slow in the face of a rapidly changing environment, it won't adapt quickly to adverse situations and will soon become just another entry on the roster of the extinct. If it mutates too rapidly, it will evolve out of its ecological niche and either starve, get eaten, or otherwise become a square peg in a round hole.* This might suggest that there's an optimum, average rate of

---

**Recent research by biochemists at Harvard University suggests that if more than a half dozen mutations in a genome occur per generation, the entire species faces extinction.*

mutation that ensures long-term survival in a cosmic context like Earth's. However, recent research by so-called computational biologists—researchers who study life by manipulating computer code rather than messy protoplasm—find that mutation rates are rather lower than optimal for protracted survival because, in the short term, a mutation is usually bad news for the individual. In a sense, mutation rates are similar to overly rigid union rules: They protect workers' jobs in the short term, but could be detrimental to the long-term survival of a business if the competitive landscape really changes.

Even though the work of the computational biologists indicates that the feedback mechanism governing mutations isn't perfectly optimized for long-term survival, Darwinian evolution still seems to be good enough, because no "perfect storm" of catastrophe—no onetime event of overwhelming calamity—has ever managed to fell the tree of life and sterilize our planet. Opossums have been doing their marsupial thing for 50 million years, which means they've seen ice age glaciations come and go. The oldest ancestors of horseshoe crabs date back 400 million years, and some sulfur-eating bacteria near hot ocean vents are probably little changed from their several-billion-year-old progenitors. This story is worth mulling over at your next book club when someone dolefully explains to you that this time, for real, we're on the edge of destruction. The historical record says that's a bad bet.

HOW TO START LIFE

Life is as durable as Christmas fruitcake. But some biologists doubt that aliens exist not because of some presumed fragility of life once established, but because of its uncertain beginnings. How likely is it that life will arise in the first place? And even if it does—even if tens of billions of worlds in our galaxy are fuzzy with pond scum—does that guarantee that some of this living matter will evolve to sentience, science, and civilization?

Let's take this one step at a time. The first question—what's the probability that life will arise?—might be answered either by unraveling how it began on Earth or else by finding it on another world. The former would give us insight into whether the process is commonplace or not; the latter would immediately show us that it clearly works—*whatever* the process—in more than one locale, and is therefore not anomalous.

Finding clues to life's earliest moments on Earth is tougher than overcooked road kill. The trail is not only cold (after all, terrestrial life goes back billions of years), but the first instances of life—soft, fragile gels—don't make good fossils. So it's hard to work from direct evidence. This difficulty has hardly slowed the theorists, however. They've been endlessly inventive in speculating how the stuff of early Earth—sloshing around on a watery planet with a thick atmosphere—might somehow self-organize into replicating life.

The seminal experiment in this field took place at the University of Chicago in 1953, when graduate student Stanley Miller defied the expectations of his adviser, chemist Harold Urey, and mimicked Earth's ocean and atmosphere of four billion years ago in laboratory glassware. Miller filled several tubes and flasks with water vapor, ammonia, nitrogen, and hydrogen—the ingredients (it was thought) of our atmosphere in those distant days. Two tungsten electrodes were hot-wired to produce sparks within the flasks, simulating lightning. Within a week or two, this bottled environment cooked up a brown, tarry goo containing 13 different amino acids, sugars, lipids, and other organic gunk. Charles Darwin had written in 1871 about life arising spontaneously from the chemical soup of "some warm little pond." Miller's remarkable experiment suggested that going from chemistry to creatures might require no more than tickling ponds with an occasional electrical storm.

In the past half century, Miller's experiment has been refined and improved. In a particularly well-known follow-up, astronomer Carl Sagan, together with a Harvard University colleague, Bishun Khare, repeated Miller's lab work in the early 1970s, only they changed the mixture of gases to methane, ammonia, water, and hydrogen sulfide in an attempt to better mimic the conditions of early Earth (as scientists had come to believe they were). They also used long-wavelength ultraviolet light (it would give you sunburn), rather than electric sparks to provoke their bubbling cauldron of ingredients into chemical action. The result was to produce amino acids yet again, as well as adenosine triphosphate (ATP), a compound essential for metabolism.

In addition, a recent reanalysis of Stanley Miller's original work shows that the Chicago chemist had actually performed an important variation of his experiment that he scarcely mentioned. He would occasionally inject steam into his glassware to mimic a volcanic explosion. Doing so, it turns out, produces an even greater roster of organic material, suggesting that volcanic eruptions could have been a particularly rich source of life's ingredients billions of years ago.

Biologists have found other ways to stock primitive life's larder. One imaginative theorist suggested that organic compounds were made when the higher tides of early Earth vigorously sorted grains of material on prehistoric beaches.* As a result, radioactive elements, such as uranium or thorium,

---

* *The tides were once higher because the moon was closer to our planet.*

were lined up on the tidal flats by wave action. While not concentrated enough to produce an inconvenient nuclear explosion, low-level fission of these radioactive lines-in-the-sand released enough energy to stimulate the formation of amino acids and the other simple ingredients of life.

The bottom line of all this work is the same: Brewing up some of biology's fundamental ingredients is as easy as remedial kindergarten, demanding only air, water, and energy.

Unfortunately, a daunting problem remains. While amino acids, sugars, and other easily concocted chemicals are certainly among biology's building blocks, discovering ways they can be made in a pond, on a beach, or anywhere else doesn't prove that the emergence of life was a straightforward process. What you really want is a Miller-type experiment (or something like it) that can produce *proteins*. Or DNA and RNA—the nucleic acids that life uses for reproduction and metabolism. But we don't know how to conduct that experiment. Manufacturing building blocks is easy, but constructing the edifice of life is a lot tougher.

This sobering fact was famously spotlighted when British astrophysicist Fred Hoyle noted that nucleic acids are so incredibly complex, he doubted you could *ever* cook them up as a random chemical process in the planetwide flask of early Earth. In 1983, he visualized the odds against this happening with the following metaphor:

A junkyard contains all the bits and pieces of a Boeing 747, dismembered and in disarray. A whirlwind happens to blow through the yard. What is the chance that after its passage a fully assembled 747, ready to fly, will be found standing there?*

**While Hoyle meant well, his remark has principally served the cause of*

As anyone who has lived next to a junkyard can attest, that doesn't happen very often. But some have argued that this is a false analogy, because biological evolution wouldn't have unfolded in the one-shot, random fashion of a junkyard tornado flinging parts together.

Well, then, how might it have worked? There's no lack of opinion on the subject, and many "self-assembly" scenarios have been proposed.

For example, geophysicist Louis Lerman would replace Darwin's primordial soup with primordial bubbles. As anyone who has churned their bathwater is aware, there are always molecules around that like to form bubbles. Lerman figures that some organically important molecules in the primordial oceans were inevitably caught in bubbles, rose to the surface in natural mini-balloons, and then burst. This would expel the molecules into the atmosphere, where a reaction akin to the Miller experiment would cook them into yet more interesting compounds. Many of these complex compounds would, of course, be washed back into the oceans, to be bubbled up again for a new cycle of chemistry. This enrichment process could eventually produce molecules that use energy and reproduce. Indeed, their bubble enclosures could function as primitive cell walls.

Another suggestion is that quantum mechanics might have had its spooky hand in genesis. In the same way that quantum computers can do many calculations simultaneously by taking advantage of the wave properties of matter, maybe the early chemistry of Earth was also multitasking—doing myriad experiments simultaneously, until it stumbled upon

---

creationists. They use it as an argument that life required the direct intervention of an outside agent.

Life on Earth may have begun in hot undersea ocean vents.

a molecule that could reproduce (and immediately start taking over the world). This would be like sending zillions of tornadoes through a like number of junkyards, over and over and over (and keeping any useful subassemblies, too). This is why some scientists suggest that, someway and somehow, quantum mechanics must have been essential to life's beginnings. With so many "experiments," the emergence of a jetliner would not be particularly miraculous.

While the role of quantum mechanics is intriguing and the bubble theory has a lovely effervescence, the favored horse in the origin-of-life sweepstakes these days is more prosaic: just the dirty, messy chemical reactions taking place in hydrothermal vents in the briny deep. These vents are open wounds in the sea bottom—rifts opened by tectonic activity as new seafloor is produced by the welling up of Earth's molten innards. Here, water that has been sucked down into the hot rock of Earth's mantle and fortified with metals and minerals gurgles back up into the cold seawater above. This is potent soup with plenty of food for life, but there's more to it than that. When the gases that accompany these underwater geysers, such as hydrogen sulfide (think rotten eggs), carbon dioxide, or hydrogen bubble through iron and sulfur compounds near the vent, they produce some of the organics of life, just as in Miller's experiment. All the ingredients for cooking up life are present, and the high temperatures ensure that the chemistry is fast and sharp.

The details are debated, but the fundamentals of the undersea vent recipe are simple: chemicals from Earth's rocks and energy from Earth's heat. The heat is considerable: Water debouching from the vents can reach a scalding 750°F (399°C), which is 400°F (204°C) hotter than french fry oil. In addition, life starting deep in the oceans would be shielded both from the nasty ultraviolet above (very little protective ozone was present at the time) as well as incoming asteroids that would pummel any nascent protoplasm. Lending encouragement to this scenario is recent evidence found in the forbidding ocean depths near Greenland proving that plate tectonics was well under way 3.8 billion years ago. So hydrothermal vents surely appeared on the scene very early.* Note to Darwin: boiling, bubbling soup rather than warm pond water.

**This hypothesis, of course, begs the question of whether we can expect other planets to have plate tectonics, and therefore the sort of underwater vents that could have been the incubator for life on Earth. Astronomers have found abundant evidence that Mars had such continental movement, although there's controversy about whether Venus ever did. Nonetheless, the red planet example shows that tectonics is not massively infrequent, but a geologic restlessness that may be a feature of many rocky worlds.*

In the interest of full disclosure, I mention one other occasionally fashionable hypothesis for life's first steps on Earth. Namely, that there weren't any. Instead, life on our planet was actually a transplant, an immigrant from another world.

Imagine, as an example, that chunks of rock were routinely kicked off Mars, catapulted into space by the incessant bombardment of incoming meteors. This sort of inadvertent rock launch could conceivably have sent Martian microbes (if there were any) along for the ride. Approximately 5 percent of all the rocks ejected from Mars will eventually hit the Earth, and lab experiments have shown that hitchhiking microbes could survive the trip. So maybe life didn't begin on Earth at all, and we are simply descendants of an early Martian infection.

Regrettably, this idea, provocatively called panspermia, doesn't help us decide if planets around other stars might carry life, because it doesn't give us any insight into how life began. It just makes the problem of biology's origins someone *else's* problem. However, panspermia's importance would change if life could survive in rocks that travel not just between adjacent planets but between the stars. If interstellar infection is possible, just a few points of genesis—or even one—might conceivably seed the entire galaxy. So life's beginnings could be highly improbable, but life's distribution could be widespread. In essence, the "biosphere" would extend over light-years.

While panspermia has its adherents, astrobiologists are skeptical about its effectiveness over such long distances. Yes, microbes can survive a jaunt between planets in a solar system—Mars to Earth, for example. Experiments in which some hardy single-celled creatures are subjected to the brutal conditions of space have shown this. But withstanding the hazards (mostly radiation and desiccation) of a million-year ride between the stars seems too much to ask of any rock-riding organism.

IT'S ALIVE!

This list of theories as to how biology might have begun on Earth, abbreviated in consideration of the reader's finite life span, should convince you that we still haven't figured out what really happened. But we could try a completely different approach to the problem—we could produce life in the lab. Doing so would be a "demonstration of feasibility." After all, if we can foment metabolism in a petri dish during the few years of a research grant, then we'd feel reasonably confident that nature could do the same in the heaving cauldron of an entire planet, given a few millions or billions of years (and no research grant).

Numerous scientists, in a modern reprise of Victor Frankenstein, are trying to do this. Some begin with only the chemicals of life, while others (emulating Doctor F.) start with existing biological parts: bits of DNA. The latter is cheating, of course, since it circumvents the thorny problem of constructing DNA. But this jump-start approach promises a faster road to artificial life.*

You may be reading about making synthetic creatures soon. But as of this moment, and despite the promise and the auspicious press releases, brand-new life hasn't stumbled out of anyone's basement laboratory.

This leaves us with the third approach to estimating the chances for biology on other worlds: Forget the labs, the uncomfortable fieldwork, and the deep-sea submersibles.

**It should be pointed out that artificial life—in particular, engineered life—might have enormous practical applications. Despite the scary scenarios that have undoubtedly crossed your mind, this development would actually improve your existence. Imagine microbes that could produce hydrogen fuel for your car, or synthetic "guinea pigs"—organisms that would allow a far faster, and more humane, test of drugs to combat cancer and other illnesses.*

Bring in the rockets. Just go find life elsewhere. On Mars, maybe, or one of the other half dozen or so worlds in our solar system where we believe that liquid water might be present. That hands-on approach eats up lots of NASA's budget.

In a sense, finding extraterrestrials nearby has already happened, although the results have been equivocal. Earlier, I described the Viking experiments of the 1970s in which we sent small biological laboratories to the red planet in the frustrated hope that they might find life. Two decades later, in a weird reversal, a new discovery raised a lot of hoopla about the possibility that Martian life had come to Earth.

The excitement concerned a meteorite with the descriptive name ALH 84001. This hunk of rock, the size of a large baked potato, had been kicked off the surface of Mars 16 million years ago by an errant meteor. It wandered the empty spaces of the inner solar system until, by chance, it slammed into the Earth during the last ice age. Rather than dunking into an ocean or falling in with an anonymous crowd of other rocks somewhere in the outback, ALH 84001 had the good fortune to land in Antarctica. In a sea of white, the conspicuous object eventually caught the eye of a NASA researcher. He, in turn, fished it off the ice and brought it back to the lab. In 1996, it became a front-page story for three days running in the *New York Times*. Why? Because NASA scientists and a chemist at Stanford University opened up this rock and found what they claimed was evidence for Martians—very small, very dead microbial Martians.

Before the news of the entombed microbes could be heralded in a press release, the story was leaked. I first became aware of it when Michelle Murray, an administrative assistant at the SETI Institute, walked into a meeting and handed me a note. It said, "Call CNN."

Wondering why, I heaved out of my chair, left my colleagues, and made the call. The San Francisco CNN reporter, Don Knapp, didn't bother with niceties.

"Seth, what do you think of this story of life in a Martian meteorite?" I responded that I didn't think anything of it, because I didn't know anything about it. I recited the names of a few Mars specialists at nearby NASA Ames Research Center and returned to the meeting.

I wasn't there long. Murray soon popped into the room again with another note, and I was soon back on the phone with CNN. "The NASA guys are under embargo," I was told. "Can you come up to the studio tonight and talk about this?"

While this was happening, a producer for ABC's *Nightline* was inveigling Frank Drake to do the same. Within the hour, Drake and I were in caucus, trying to figure out what the heck we were going to say on national television about a story we knew nothing about. Fortunately, the *Nightline* producer somehow managed to get a prepublication copy of the still embargoed paper (how he did this remains mysterious to this day). He faxed it to the SETI Institute, and we crammed for a few hours. That evening I spent seven minutes bobbing and weaving with reporters in Atlanta about the significance of finding fossilized microbes in a rock from another world.

It was a big story—and my mom thought I looked good on TV—but its import eventually shriveled. In the past dozen years a small horde of astrobiologists have examined the rock, and few of them are convinced that ALH 84001 (or any of the other three dozen meteorites known to come from Mars) make a compelling case for biology on the red planet, either dead or alive. The specialists are still hashing this matter out, but the word at astrobiology mixers is that we won't really know if Mars has or had life until we either go there ourselves, or send some landers to the red planet that can pick out some good rocks to examine—rather than relying on the random samples sent our way by meteors.

It's remarkable that, 130 years after Schiaparelli charted his canali, Mars still manages to intrude into any polite conversation about extraterrestrial life. Mars is no longer the only game in town, however. Among the other sirens on the shores of the solar system are Europa, Ganymede, Callisto, Titan, and Enceladus. In a revelation that would startle your grandaddy's astronomer, researchers have learned that these moons of Jupiter and Saturn might be sufficiently stretched and squeezed by the gravity of their host planets to melt the ice near their surfaces, producing hidden oceans of liquid water. Consequently, any of these small, strange worlds could support life.

In the next three decades, our spacecraft will investigate in greater detail some of these moons. How will they do that? NASA is considering a mission to Europa that would land on the surface, melt a hole through an estimated 10 miles (16 kilometers) of hard ice, and drop probes into the hidden water below to hunt for Europans.

That water is pitch-black—all sunlight is blocked by the ice above—so it's unlikely that large, complex life will be found. There's simply too little energy available for much of a food chain. If Europans exist, they'll probably resemble some of the extremophiles described earlier.

Similarly, researchers have considered the possibility of life on Saturn's moon Enceladus. When close-up views of this ice ball were made by NASA's Cassini spacecraft in 2005, scientists were stunned to find it spouting a geyser of icy water. Could life lie somewhere within? It's not impossible, as demonstrated by the earthly extremophiles that can survive in conditions that mimic the cold, subterranean roots of Enceladus's spectacular spouter.

Single-celled extremophiles may be tantalizing examples of the kind of life we might find elsewhere, but let's face it, pond scum doesn't qualify as intelligent life. The conditions on these nearby satellites are just not salubrious enough for highly complex, thinking beings. However, calculations suggest that moons far larger than Europa or Enceladus—moons the size of Earth—could be tidally heated at a hundred times the rate of the Jovian moons. If terrestrial-size moons exist around some of the large extrasolar planets that astronomers continue to uncover, they might be sufficiently Earth-like at their surfaces to permit the emergence of a diverse biota, possibly including intelligent life. Moon dwellers may yet exist—on another planet's moon.

However, and despite the manifold possibilities, we still haven't found any of this putative extraterrestrial life. We haven't unraveled how biology began on Earth either, nor have we produced life in the lab. In tabulating our attempts to understand whether life is a long shot or a safe bet, we've scored zero for three so far.

But there's one fact about life that's neither theoretical nor awaiting new experiment. We can say for sure that our planet was gilded with protoplasm from a very early age. That could be telling us something important.

LIFE: NOT A BIG PROJECT

Think about ancient life. Perhaps you picture dinosaurs or, if you have a fetish for segmented sea life, maybe trilobites. But in truth, these life-forms—nicely stacked and racked at the

Stromatolite fossil in 3.4-billion-year-old Australian rock

local natural history museum—are recent arrivals: They're yesterday's fauna. Imagine going back a lot further, to a time when Earth was only a billion years old, one-fourth its present age. That world featured a gray expanse of tepid ocean broken only by scattered chains of volcanoes. The atmosphere was oppressively warm and glutted with carbon dioxide. Oxygen levels were so low you would suffocate in seconds. And yet this strange and hostile place—where even continents were but a distant dream—was already fringed with life.

We know this because we've found its mummified remains. In northwest Australia there's a bit of arid real estate known as the Pilbara hills. Thanks to the vagaries of geologic history, the Pilbara sports the globe's best-preserved early sedimentary rock. From radioactive dating experiments, we know (very accurately) that this rock is 3.4 billion years old.

Biologists and geologists have pored over the Pilbara for years, combing the ancient strata for clues to Earth's primordial history. In particular, they've focused on some bumpy, muffin-like features found within the sedimentary layers. They're now convinced that these conical and dome-shaped structures are fossilized stromatolites—homes built as ancient algae mats in shallow ocean pools. Stromatolites are analogous to coral reefs: relatively big, communal homes patiently constructed by tiny occupants. Indeed, we know a lot about stromatolites because they're still around; the best examples can be seen in the tidal waters of Australia's Shark Bay and the steamy pools of Wyoming's Yellowstone National Park. These moundlike artifacts of microscopic life are relatively uncommon today, but they were widespread billions of years ago, before single-celled "grazers" evolved that could feed on the builders.

Stromatolites may not be your favorite fossil, but they tell a profoundly important story. It seems all but certain that 3.4 billion years ago, the scabrous red landscape of the Pilbara was a flourishing ecosystem, riddled with life.

And how many years earlier *than that* did life begun? We don't know. But keep in mind that, although a billion years passed before stromatolites formed, Earth was an insufferable place for nearly all that time. Our planet was a metal duck in a cosmic shooting gallery. A lethal hail of rocks from space, leftovers from the violent birth of the solar system, routinely pummeled Earth's hot landscape. By carefully studying the moon's craters (the moon, after all, was also a target for these unguided missiles), we know that the storm of rocks abated about 3.8 billion years ago, at the end of what's picturesquely called the "late heavy bombardment." That was when the cosmic weather cleared. Our Earth, replete with possibility, was able to stumble into its florid destiny without the constant havoc of projectiles from space. And shortly thereafter, life began.

Or maybe not. As rapid as this process may seem, life may have been under way hundreds of millions of years earlier, even while asteroids were still hammering the planet. The evidence for this head start comes from zircons, which are crystals of zirconium silicate about the size of a pin. These miniature crystals, long recognized as geology's best time capsules, form when hot magma from Earth's mantle reaches the surface and cools. Zircons are tougher than vulcanized brisket, able to withstand the inevitable weathering, heating, and crushing that eventually destroys nearly every other type of rock. Recent analysis of small bits of diamond found in some very old zircons—microscopic gems trapped like flies in amber—has revealed a composition suggesting that their elemental carbon was once part of some living entity. This evidence, although still ambiguous, implies that life was simmering in Earth's waters only 400 million years after our planet came into existence.

When, exactly, life started on our planet is still debated. Whether it was 3.4 or 4.2 billion years ago, the take-home message is this: If life arose so shortly after our world was born, a period of time that's no more than a sigh in a planet's history, then it could scarcely be a wildly improbable event. Nature didn't waste much time starting its terrestrial biology project. So how hard could it be? Life is likely to happen often.

A conceivable objection to this idea is the argument that, yes, Earth cooked up life quickly, but that's only because our kitchen was special. Can we assume that the ingredients of life that were present on Earth would be available on other planets around other stars? In other words, was Earth uniquely favored in some way with the makings of life?

The evidence says no. The elements of terrestrial life—carbon, oxygen, nitrogen, hydrogen, phosphorus—are found on all the rocky worlds of our solar system and, one can safely assume, on similar-size planets in other solar systems. So although we don't yet know how life arose on our planet, we do know that there was an abundance of raw material—indeed, the material wasn't even that raw. And those ingredients were made up in large batches during the apparently quite ordinary Mixmaster era of our solar system's formation. So while the recipes for life elsewhere might be different, the larder could be very similar. That isn't to say that life on other worlds would be biochemically identical to terrestrial life, of course. Your local taqueria proves that many different dishes can be made with the same storehouse of ingredients. But it does suggest that most planetary systems will at least have a shot at serving up life.

HIGH-IQ LIFE

All of this discussion is a tentative but promising argument for ET's existence. Indeed, even biologists—who love to groan about how astronomers and other nonexperts play fast and loose with the question of life's existence elsewhere—generally agree with the optimistic assumption that life has appeared on many, many worlds. After all, if life is just one more natural product of the cosmos, then biology, like physics, is universal. But many biologists draw the line when you take one more step, that not only is life likely to be widespread, so is *intelligent* life. Even if our assumption that billions of worlds in our own galaxy are frothing with life, we might still be alone, at least in the sense that there might be no one to talk to. Intelligence might be rare; indeed, *extraordinarily* rare.

Having studied life's history in eighth grade, you may wonder at this assertion. After all, the story seems pretty straightforward in retrospect. Three to four billion years ago, life started in some watery wastes. After a few billion years, this quiet microscopic life evolved multicellular forms. Roughly a half billion years ago, the Cambrian explosion occurred, and suddenly Earth was literally crawling with critters. Trilobites appeared, but so did ferns, fish, and frogs. The dinosaurs were the planet's superstars for an impressive 150 million years (without, it should be noted, leaving a single instance of noble architecture behind). Their long run came to a bad end when they, along with about two-thirds of all other species, were rudely exterminated by an errant asteroid. But the mammals picked up where the dinos left off, and eventually produced us.

Looking back, it all seems inevitable.

Of course it does. And of course it wasn't. *Homo sapiens* is the consequence of many forks in the evolutionary road, and if any one of the other roads had been taken, our species wouldn't have appeared. Geophysicist Andrew Watson, in England, has made probability estimates of the major steps in the evolution of sentient life on Earth, and figures it's a 1-in-10,000 shot, even for planets that are suitable for life. Or consider this simple argument: If the five-mile-wide (eight-kilometer) rock that hit what is now southeastern Mexico on an unassuming day 65 million years ago had arrived 20 minutes earlier, it would have sailed by our planet without notice or incident. Your neighbors would be obnoxious dinosaurs, not obnoxious hominids.

That makes it sound like sentience on Earth was really a long shot, and we were just lucky to hit the evolutionary jackpot. However, the question is not whether *Homo sapiens* was likely to arise, but whether functionally similar intelligence would do so. Maybe the dinosaurs would eventually have evolved humanlike intelligence and developed a science-based technology just as we have. (Presumably these sauropods would look back on the great Permian extinction and wonder whether *"Dino sapiens"* could have evolved if *that* catastrophe hadn't happened.)

A quarter century ago, Canadian paleontologist Dale Russell speculated that a dinosaur known to have a relatively large brain, a troodontid (known at the time as a *Stenonychosaurus*), could have produced sentient descendants.

Russell cooperated with a taxidermist, Ron Sequin, to produce an appealing model of *Stenonychosaurus*'s putative successor: a nerdy, highly evolved "dinosauroid." It looks human, other than the three-finger hands, scaly skin, and lack of teeth, and has been pictured in books about as often as many real dinosaurs. But even Russell admitted that his hypothesized dino descendant might be too anthropomorphic. More recently, others have stated that the dinosaurs, had they survived, would have retained their horizontal body posture, looking more like strange birds than strange humans.

But what they'd look like is less important than whether they would have reached our level of intelligence. The troodontid, while certainly near the head of his dinosaur class, had the IQ of a raccoon. If humans were wiped out tomorrow, do you imagine that raccoon descendants tens of millions of years hence would be driving cars and sitting on the couch, wolfing down nachos and avidly watching sitcoms? That's a highly speculative call. I once asked Niles Eldridge, a paleontologist at New York's American Museum of Natural History, whether the dinosaurs, if spared oblivion, might have become brainy. His response betrayed a hint of doubt: "The dinosaurs had 150 million years to get smart, and didn't," Eldridge said. "So what would another 65 million years have done for them?"

Whether the monsters of the Mesozoic would have made the cerebral grade is an interesting sideshow to the main research attraction—is intelligence truly inevitable on worlds with rich biota? Similarly to the earlier question on the probability of life, one might find an answer in two ways. Either we could determine how intelligence arose on Earth, thereby establishing whether the mechanism is likely to be common on other worlds, or we could find intelligence elsewhere.

Do we know why our terrestrial ancestors got smart? Have we deciphered the evolutionary mechanism that caused our predecessors of three million years ago—with their 1.1-pound (0.5-kilogram) brains — to rapidly develop the 3.3-pound (1.5-kilogram) brains that qualify them to be *Homo sapiens?*

We don't and we haven't, although scientists will frequently take recourse in a mechanism called convergent evolution. This is the observed fact that similar environments will eventually produce similar adaptations, even in unrelated organisms. A canonical example is the streamlining of ocean predators. The torpedo-shaped bodily structure of barracuda and dolphins is simply good engineering for a species whose survival depends on getting through the water quickly. These species are only distantly related; their similar silhouettes are the consequence of convergent evolution in response to their aqueous environment.

Australian astrophysicist Charley Lineweaver says streamlining is one thing, but he doubts whether convergent evolution is a powerful enough engine to churn out intelligence. He points out that, yes, our ancestors arose in Africa for some as-yet-unspecified reason. But a half dozen other landmasses—effectively isolated for tens or hundreds of millions of years—are doing their own convergent evolution experiments. And they haven't produced anything remotely as quick-witted as humans. These other regions include New Zealand, Madagascar, Australia, and even large landmasses like North and South America. Lineweaver has studied the smartest creatures to arise in these other biomes and finds them all distant seconds to us.

Frank Drake has responded to Lineweaver's argument by saying, "True, they're not at our level. Yet." His point is that whichever species first makes the intellectual grade will look around and say, "Hey, there aren't any others!" So maybe there's nothing remarkable in that fact.

Other researchers, too, take a more optimistic view than Lineweaver on the chances for cleverness. They've identified several reasons why a range of animals is pushing toward greater intelligence. Octopuses, simians, and some birds (like parrots) are among the creatures that display above-average intelligence, and they're not all closely related. Challenging environments (ice ages, for instance) tend to selectively dispense with the less adaptable individuals in a given species. But social environments might be even more important in favoring the intellectually endowed. For highly social animals, being smarter than the fellow members of your species clearly carries a benefit. If you can anticipate the behavior of other group members, clearly you have an advantage both in getting dinner and getting a mate. Spotted hyenas, like primates, are both social and smart. A study by behavioral biologist Lori Marino of highly social dolphins and toothed whales from the past 50 million years shows that some (but not all) species have dramatically increased their intelligence, at least as gauged by the ratio of brain to body mass.

So that's one explanation for your high IQ: your ancestors' social life. Another activity that might have spurred intelligence is building things. Bowerbirds and beavers are both clever, and for both, construction projects are a big part of life. Bowerbirds build their bowers (highly decorated nests with ornamental "front lawns") to attract mates, and beavers build their dams to corral fish. Cooperation and ingenuity are rewarded in such lifestyles.

Anthropologist Craig Stanford has suggested that our penchant for meat accounts for our intellectual talents. He argues that the social subtleties necessary to negotiate over steaks—a scarce commodity for our ancestors—might have a competitive advantage. The smartest negotiators would get the best meals, and presumably so would their offspring.

But no matter what the specifics, the fact that several species have increased their cognitive abilities suggests to at least some biologists that convergent evolution really is at work, adapting to the niche market for animals that, in the end, are smarter than your average bear. The mechanisms that produce intelligence are still unclear, and proof of what propelled our predecessors toward brainy behavior is still lacking. Maybe we'll short-circuit this search for the history of our genius by uncovering intelligence elsewhere in the galaxy. That would tell us at a stroke that such a development is clearly not enormously improbable. And, of course, finding intelligence elsewhere is the goal of our SETI experiments.

DO SCIENTISTS REALLY BELIEVE IN ALIENS?

Sure, SETI researchers hope that intelligent life is widespread. If they didn't feel that the chance for cosmic consociates is good, they'd choose another line of work. That would at least free them from endless anxiety about funding.

But what about other members of the scientific community? Are they equally sanguine?

Most, but not all. Astronomers see the enormous plentitude of the universe, and say that it would be offensively self-centered to imagine that what has happened on Earth has happened *only* on Earth. Physicists look at the laws of nature, find that they're universal, and conclude that if sentience could happen once, it has surely arisen many times.

But, and as we've noted, biologists are not always on this same sunny page. Many of them concede that life might be commonplace. After all, unlike warp drive or time travel—both of which would require us to throw out a lot of existing physics in order to be true—supposing extraterrestrial life is only a matter of extension, not of radical revision. The research of the past two dozen years has shown that literally billions of planets might have niches that would support at least the kind of stoic life represented by Earth's extremophiles. And that's just the tally in our own galaxy.

Yet more than a few biologists are less optimistic that intelligence will be routine and rampant. They point out the discouraging fact that, as a biological trait, intelligence is not progressive. Members of a species don't automatically become more intelligent as the generations go by, any more than they automatically get bigger or smaller.

But some do, and because of a simple statistical effect, that circumstance might be enough to guarantee the emergence of sentience in a world in which the total number of species keeps increasing.

On Earth, this increase in diversity has been going on for billions of years. Wipeouts by asteroids or other natural disasters have occasionally culled large numbers of species and reduced the total, but the overall trend in diversity has been upward. More species than ever now reside on Earth.

How does this help? Imagine that you could somehow give IQ tests to a few dozen different animals snagged from the capacious (and redolent) hold of Noah's ark. (To simplify matters, let's say you choose only mammals, because, after all, you're mammalocentric.) Most of these furry test takers would rack up intelligence scores somewhere near animal average, and a few would be somewhat smarter or dumber. If you bothered to make a histogram of these scores—a chart of how many scored at each IQ level—you'd find that it would approximate a bell-shaped curve, with the peak of the "bell" near the average. If you sought out the brightest beast in this small sample, it would likely have an IQ that's maybe two or three times the average. Smarter than your average bear, perhaps, but not *that* much smarter.

Now do the experiment again, but this time test *all* mammal species—all 45,000 of them. Once again you'll get a bell-shaped distribution, with the overwhelming majority scoring near the bump of the bell. In other words, most animals will have IQs near the average. But the cleverest species is likely to be at least twice as smart as the cleverest critter in your smaller sample. It will be far out on the high-IQ side of the curve (at the end of the "tail," in math-speak).

Indeed, that extraordinary beast might look around, realize that it beats the intellectual pants off not just some, but *all* of its brethren, and wonder at the fabulous improbability that evolution could produce something as smart as itself. But this result was merely the consequence of statistics. When the sample size gets larger, more species will be in the outlying "tails" of the distribution. The cleverest will be substantially

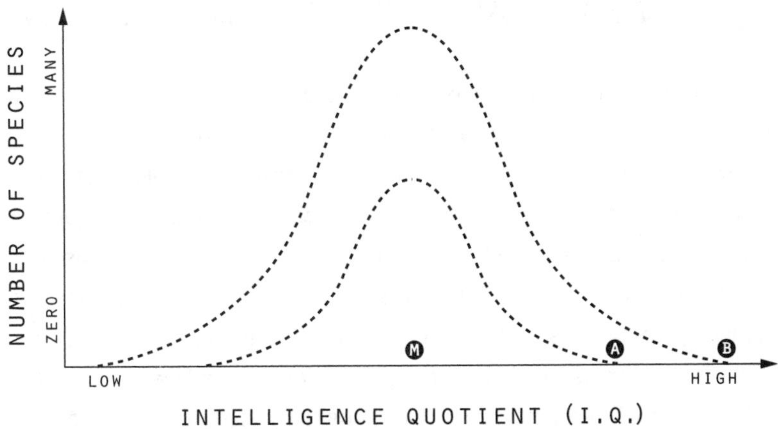

How a greater number of species can inevitably ratchet up the IQ of their smartest member: The higher bell-shaped curve, which represents the spread of intelligence for a larger population, includes one species at IQ B. The smaller population only reaches the lesser maximum A. Both populations have the same average intelligence, M.

brighter than when the number of animal species is smaller.

While this argument seems straightforward, the research community still engages in endless discussion about the probability of sentient life eventually emerging on worlds that have made the step from dead dirt to living pond scum. Will intelligences that are functionally equivalent to our own be a frequent product in such biota or not? Despite our example, the arguments pro and con remain tentative and inconclusive. Possibly the only way to end the debate will be to find sentient beings on a still unknown planet, far, far away.

But there are some people—well, in fact, a *lot* of people—who think that SETI is wasting time and money spinning its telescopes. These folks maintain that the extraterrestrials are not light-years away, sequestered around an unknown star. They're here, in our airspace, and occasionally in our bedrooms.

SETI'S FAMOUS FORMULATION: THE DRAKE EQUATION

At a celebration of Frank Drake's 70th birthday, physicist Philip Morrison called it "the second most famous equation in the world."

He was referring to a simple string of terms that Drake first wrote down in 1961 to do something quite immodest: Estimate the number of alien civilizations in our galaxy. Mind you, they couldn't be just *any* aliens. The Drake equation was designed to tally worlds that had the technological smarts, not to mention the motivation, to pump signals into space, signals our telescopes could detect if only we'd make the right choice of frequency, direction, sensitivity level, and a handful of other technical options.

The equation was intended to be an organizational tool—an agenda, if you will—for a three-day conference that the U.S. National Academy of Sciences had asked Drake to organize shortly after his pioneering SETI search, Project Ozma. The formula wasn't a foray into new mathematical territory, merely a straightforward concatenation of first-order terms. It wasn't even the first time that such a calculation had been considered. Only two years earlier, American astronomer Harlow Shapley had surveyed the cosmic landscape and decided that extraterrestrial intelligence must be a common occurrence. Shapley reckoned how frequently life might appear on other planets, and did so by multiplying a short series of factors: What fraction of stars hosted planets, what fraction of those were warm enough for life, what fraction of those were sufficiently large to hold onto their atmospheres, etc. The product of these factors—Shapley's bottom line—was that one in a trillion star systems might host a planet with life, an estimate so pessimistic that it implies that the majority of galaxies are as sterile as a hot poker.

Drake's formula was both more specific and better constructed. He intended it for our galaxy only, and included some terms specifically relevant to the sentient life that Shapley had vaguely predicted would be out there. There are several variations of the equation, but Drake himself usually writes it as

$$N = R_* \, f_p \, n_e \, f_l \, f_i \, f_c \, L$$

Before I delve into the necessary tedium of elaborating each of these factors, allow me to point out the formula's simple underpinning logic. Imagine you want to estimate the number of married folk in the U.S. One way to do so is by multiplying the annual number of American marriages times the average number of years people stay bonded together in matrimonial bliss. It's simple—the number of people in a particular situation equals the rate of entry into that situation times the average length of time spent there.

The Drake equation multiplies the number of sentient societies that are born each year in our galaxy—the first six terms of the equation—times L, the average number of years they remain in *technological* bliss, belching out signals for our benefit and erudition. The result is N, the estimated number of galactic civilizations amenable to discovery by our SETI experiments.

Having painted the big picture, let's briefly consider the individual terms.

R_* is the number of stars born annually in the Milky Way that could host life-bearing worlds. This factor can easily be estimated, since it's based on astronomical measurement. The answer is one or two dozen stars per year.

f_p is the fraction of such stars that actually have planets. Thanks, once again, to the sedulous efforts of astronomers, we have some firm knowledge here as well. As noted elsewhere in this chapter, Geoff Marcy believes that between one-half and three-quarters of all stars will eventually prove to have planets.

n_e is the number of worlds per solar system that have the conditions for life. In 1961, the subscript e was a reference to Earth, the assumption being that only analogs to our own world would be amenable to biology. Today we know that certain tough organisms (extremophiles) might be able to survive on some of the moons of Jupiter and Saturn—worlds, it turns out, that might house liquid water. These revelations suggest that, whatever we used to believe was a good value for n_e, it should be raised a bit now. A conservative guess here would be "a few," although it also has to be said that we currently know of *no* other worlds with conditions similar to Earth's.

f_l is the fraction of worlds boasting the conditions for life that actually *have* life. We're now entering dark territory because we know, to date, only one example of such a world: our own. And you can't generalize from a sample of one. On the upside, the universe seems rife with the organic compounds that underpin carbon-based life, although an abundance of raw materials doesn't guarantee a lot of product. The traditional estimate for this term is 0.1. Traditions aren't always good guides to truth, of course.

f_i is the fraction of worlds with life that eventually produce intelligent life (fill in your own definition of "intelligence"). This number is highly controversial—after all, was *Homo sapiens* (or something like him) inevitable, or merely a weird product of four billion years of convoluted evolution? In truth, while strong opinions rage about whether intelligence will emerge on worlds with life frequently or otherwise, we possess no unambiguous evidence one way or the other. Optimists peg this factor near 1, but you won't be without company if you're less sanguine.

f_c is the fraction of worlds with sentient beings that develop communicating technology (this is equated to "civilization," hence the subscript). In retrospect, it seems inevitable that once our species arose, it would eventually develop the understanding of science that would allow us to build radios and telescopes. But anthropologists can point to many societies (perhaps one should say "most societies") where such a path of progress was not taken. Still, if science is invented by any group, its utility certainly seems likely to spread to the others. For this reason, a value near 1 for this term is often adopted.

If you multiply our estimates for the first six terms above, you find that they predict the genesis of one or two technological societies each year in the Milky Way. All that remains of this algebraic activity is to estimate L, the average lifetime of such societies.

The value for L cited by Drake himself is 10,000 years. Others have made estimates considerably shorter and longer. But as everyone who dares to speculate will admit, this is the one term where our guesses are well and truly guesses. Indeed, some have said that the most important justification for SETI is to get a handle on L, and learn whether we can survive our own technology. In the interim, and taking the estimates noted here, we compute that 10,000 to 20,000 societies stud the Milky Way.

The Drake equation, despite our inability to nail down many of its factors, remains as much an organizational structure for thinking about intelligent life in the universe as it did in 1961. Many people have written Drake to suggest modifications to his famous formulation. (One item that comes to mind is the absence of any consideration of interstellar colonization. The equation assumes every technological society stays put.) But Drake has not seen the need to change anything.

"The equation has stood the test of time," he says. "I do get suggestions for additional terms, but so far all such factors can be subsumed into the original formulation."

Oh, and by the way—what's the *first* most famous equation, according to Morrison? $E=mc^2$, by Albert Einstein.

COULD THEY BE HERE?

As if the woes of this world weren't enough, aliens from outer space were messing up Ralph's life. He called me for the third time in as many months.

"I can't sleep," said the plaintive voice on the other end. "The floating energy balls make noise all night long."

"What kind of noise?" I asked.

"Sort of a *whooo-whooo.* They sway back and forth. Nonstop. Keeps me up."

According to Ralph, extraterrestrials had set up the outdoor energy balls (whatever such things are) as a deliberate annoyance. I inquired as to their size and position.

"Oh, they're about three feet across, and maybe a hundred feet up," he said.

"Well, that makes them easy to see," I ventured. "I bet these things are bothering other people too. Why don't you get together with a few neighbors, take some photos, and complain to the authorities?"

There was silence. After ten or fifteen seconds, Ralph's tentative voice returned: "I don't have much contact . . . with my neighbors."

I suspect that Ralph's problems were internal, not interstellar. But this conversation was hardly unusual. As one of a handful of scientists who actively search for evidence of cosmic habitation, I receive a lot of calls and correspondence from folks who have something to say on the subject of aliens. Many of them, like Ralph, are having difficulties with otherworldly visitors in their personal lives.

These odd exchanges are an inevitable consequence of the fact that a large number of Americans (and Europeans, Japanese, and Australians) are convinced that aliens are not only "out there," but also here on Earth, violating our airspace and our personal space. In a 2002 poll conducted by the Roper organization, 56 percent of Americans said they believed that some UFOs are alien flying machines, and an even higher percentage (72 percent) reported being convinced that the U.S. government knows about extraterrestrial visitors and is keeping mum. Other surveys made during the past several decades have produced similar results.

The dual belief that sentients are visiting Earth and that a pernicious conspiracy is masking their activities is common. Consequently, I'm not surprised that some of the vast multitude who believe this entwine their day-to-day problems with imagined interference by beings from beyond. Their complaints are often more severe than mere sleep deprivation. One afternoon I was phoned five times by a distraught woman named Carrie, residing somewhere in Saskatchewan. Carrie was desperate because her father had threatened to feed her to an alien camped out in a backyard shelter. She implored me to quickly travel a thousand miles to her rural locale and forestall this messy fate using my supposed expertise on alien behavior. In the course of Carrie's fifth call, as I silently wondered whether she made the right choice in opting to live with her dad, she confessed that the threat of being turned into alien chow was particularly burdensome given that she also had to juggle her multiple personalities.

Carrie was sincere, as are virtually all those who contact me to describe sightings, encounters, and novel theories about how we might quickly locate the extraterrestrials. Such people are not hoaxers. When a person calls up to describe a craft they've seen in the sky, I'm confident from their tone and earnestness that they've seen *something*. After all, thousands of people have. The question, of course, is what?

Could they, indeed, be visitors from outer space? By the end of the 20th century, the hope that our solar system might house intelligent life was nearly bankrupt. NASA's space program had shown Mars to be too nasty for complex life, and other nearby worlds looked even less attractive. But despite the disappointment, these revelations hardly leached the public's firm belief in extraterrestrials. That's because large-scale rocketry and our first forays to the moon, Mars, and beyond made space familiar territory. Sure, these exploits were modest by cosmic standards: Despite all the hoopla, we'd sent hardware no farther than our planetary neighbors. But it was difficult to resist the idea that bridging the far greater distances to the stars was only a matter of better technology.

Add to this the effects of Hubble Space Telescope imagery. Think back two decades, and consider how unfamiliar ordinary folk were with astronomical photos. True, the black-and-white pictures of nebulae made with the ground-based Palomar telescope and its ilk were known to astronomers both amateur and professional, but the general public remained largely unaware of such celestial images. Today, everyone is exposed to Hubble's steady stream of chromatic skyscapes.

Pictures of the cosmos have become so mainstream, so comfortable, that they've served to defang space. After all, if the Eagle Nebula can be plastered onto the side of a city bus, how can anyone imagine that it's truly inaccessible? Today's astronomical photos paint a universe that is strikingly beautiful. They fail to convey just how remote and brutally hostile the near-empty voids beyond our solar system are.

However, and without doubt, the biggest factor in recasting space as just a neighborly hunk of backyard real estate is the popularity of science fiction—in particular, the movies and TV shows in which aliens routinely barrel across the galaxy to engage us in conversation or conflict. In most stories, space is just the Wild West without the dust—a vast backdrop where stars replace saguaros, and where the bad guys are just like us, except for their obvious need of remedial plastic surgery.

The idea that space is tractable has an inevitable consequence: If far-flung beings can discourse with *Star Trek*'s Captain Kirk in the 23rd century, why couldn't their brethren meet us now, on Earth? In 1898, H. G. Wells postulated that aliens would come to our planet for the water. The Martians were just across the street, metaphorically speaking. But a century later, with rocketry a substantive enterprise, many people can easily imagine real aliens journeying to our world from some other star system. And something other than Earth's liquid assets might be seducing them.

THEY'RE UP IN THE SKY!

The UFO phenomenon—the claim that we're being visited—is either true or is an enormously widespread and durable fantasy. It's tempting to compare UFOs to other types of postulated visitors: A Gallup poll taken in 2005 showed that one-third of

Americans say they believe in spirits, presumably on tour from beyond the grave. A 2001 survey by the Scripps Howard News Service revealed that three-quarters of the populace profess the belief that angels not only exist but visit us in modern times.

Ghosts and angels are ephemeral, and can be declared to exist on the basis of faith alone. Sentient aliens, on the other hand, are presumed to come from worlds we could (at least in principle) see with our telescopes. They can be studied with the tools of science. Unlike winged seraphim or spectral ghosts, they are thoroughly corporeal, subject to experiment and observation. To date, experiments on UFOs are still lacking. But there's no absence of observation. According to reports, they make many appearances daily.

UFOs, in their modern incarnation, first flew onto the stage on a clear Tuesday afternoon. On June 24, 1947, Kenneth Arnold, an Idaho businessman working for a company that manufactured fire-control equipment, was piloting a small plane in the vicinity of Mount Rainier, Washington. On a whim, he decided to informally join the search for a downed military transport. At about three o'clock, he was distracted by bright, luminous flashes that he believed were caused by sunlight bouncing off a fleet of aircraft dozens of miles away. Arnold later told reporters he saw nine craft, flying at an altitude of roughly 9,800 feet (3,000 meters) and sliding above the volcanic topography at more than 930 miles (1,500 kilometers) an hour. He tracked them for nearly three minutes as they coasted from Mount Rainier to Mount Baker and beyond.

The unidentified flying machines were apparently neither military nor civilian craft, and their mysterious nature soon transformed them, at least in the public's mind, into something far more exotic: spaceships piloted by intruders from other worlds. This idea—that aliens might be cruising our airspace—caught on quicker than a burr on a sheep dog.

Scarcely a week passed before an event occurred near Corona, New Mexico, that would eventually eclipse in perceived importance not only Arnold's sighting, but just about every UFO incident since. On July 4, W. W. "Mac" Brazel, a Corona rancher, stumbled upon some debris on his property that—according to contemporary newspaper reports—included strips of rubber, pieces of paper, bits of wood, and some shiny foil. He picked up pieces of this weird litter and showed it to his neighbors. They were as puzzled as Brazel. Two days later during a business trip to Roswell—a larger town about 60 miles (100 kilometers) to the southeast—Brazel mentioned this bizarre discovery to the sheriff, suggesting that maybe he had picked up the remains of a flying disk. Flying disks had become a media sensation following the Kenneth Arnold story, and Brazel had heard of them.*

The sheriff contacted a Maj. Jesse A. Marcel at the nearby Roswell Army Air Field (RAAF), and Marcel sent someone back to Corona with Brazel to examine, and eventually collect, the debris. On July 8, the *Roswell Daily Record* was top-heavy with a dramatic headline: "RAAF Captures Flying Saucer on Ranch in Roswell Region."

In one week, the disks had advanced from a high-speed skitter through the Cascades to a crash in the New Mexico desert. The Roswell story quickly mutated. The next day,

---

**The newspaper reporter who wrote up Arnold's description of these flying objects misunderstood him. The reporter wrote that they had a saucer shape when in fact Arnold was trying to describe their motion, moving through the sky "like a saucer would if you skipped it across water." This misquote led to the term "flying saucer" and, one presumes, the suspicious fact that so many photographs of purported alien craft look like dinnerware.*

the *Record* reported that an Army general had pooh-poohed the original claim, and was now describing the debris as busted bits of a weather balloon, not an extraterrestrial transport. Interest faded, although Brazel, who said he frequently came upon detritus from crashed weather balloons and knew what these inflatables looked like, didn't believe the revamped story. The whole incident moldered for decades until writer Don Berliner and physicist Stanton Friedman investigated the Roswell incident, eventually turning it into a cause célèbre with their 1992 book, *Crash at Corona*. According to the two authors, honest-to-God aliens had made a serious pilot error above the Land of Enchantment and suffered a fatally rough landing. The U.S. military, for reasons that anyone can easily concoct, was covering the whole thing up. The debris was eventually taken to an undisclosed location and, if some pundits are to be believed, reverse engineered to speed our own development of high-tech products (see Reverse Engineering UFOs, page 150).

Now, mind you, others have advanced alien-free explanations for what transpired near Roswell. One, promoted in the past few years by British UFO expert Nick Redfern, is that the U.S. military in Roswell was actually testing Japanese plans developed during World War II for dropping bombs on distant targets using manned balloons. Redfern maintains that the scheme was a kamikaze mission, and the pilots weren't supposed to come back. He suggests that one of the American evaluation test balloons (including its one-man crew) came to a bad end in the desert dirt. This lighter-than-air bomb delivery system reputedly conscripted genetically defective personnel, a fact that Redfern believes may account for the testimony that some Roswell witnesses saw strange "alien bodies" at one of the crash sites.

Brig. Gen. Roger M. Ramey and Col. Thomas J. Dubose display debris from the Roswell crash, claimed to be pieces of a weather balloon.

The evidence for suicide balloons guided by genetically damaged pilots is equivocal. A more credible explanation is that the incident in New Mexico was an unintended consequence of a secret government effort to monitor Soviet nuclear tests, called Project Mogul. Its mission was to detect H-bomb tests from afar using instrumented balloon trains.

Mogul was undertaken on the basis of work conducted by an American physicist, Maurice Ewing, at the close of the Second World War. Ewing realized that there are "sound channels" in both the ocean and in the atmosphere. These are layers at which sounds can be contained and propagated without significant fading, almost as if they were in a giant pipe. There's nothing magical about this; it's entirely due to temperature differences that affect the speed of sound. In the atmosphere, the channel exists because air temperatures cool with altitude until you reach about 260,000 feet (80,000 meters), when they start to rise again (due to the ozone layer). So at roughly 9.3 miles (15 kilometers) up, a stratum of cold air serves as an efficient conduit for sound.

Ewing elaborated on this fact to devise a plan for sniffing out Soviet test explosions. His scheme was to launch long trains of balloons to the altitude of the channel, and outfit them with sensitive microphones that could detect the distant, churning noise of a mushroom cloud as it rose through the layer of cold air. In 1947, the RAAF was evaluating this scheme, and according to those involved, it was the debris from one of these experiments that Brazel found.

This explanation makes sense. But then, as now, the public found it far more reasonable to think the government was covering up an alien presence rather than defense secrets, and the "Roswell incident" became the poster child for those marketing the idea that Earth is being visited.

Soon, these pioneering UFO stories had plenty of company. By the end of the summer in 1947, nearly a thousand UFO sightings had been reported in the American media. The aliens were blitzing the skies—or so it seemed.

Perhaps because the memory of real blitzkrieg was still fresh, the military—which was already tooling up for the Cold War—took an early interest in the UFOs. It's doubtful that many Pentagon brass were inclined to think that these strange aerial phenomena were really hardware from someone else's star system, but they did have genuine concern that the spaceships might be hardware from the back side of the Iron Curtain.* By the end of the year, the first of a small series of official investigations into UFOs was launched: Project Sign (also known to insiders as Special Project HT-304, or occasionally as Project Saucer). This was strictly an Air Force investigation, conducted at Wright Field in Dayton, Ohio, and involving staff specializing in technical intelligence. Although the investigators first believed the UFOs were probably Soviet aircraft, within a year their sentiment had shifted in favor of the idea that some of the unexplained objects might be visitors from the stars. At this point the Air Force Chief of Staff, H. S. Vandenberg, apparently unhappy with this new slant, decided to cancel Project Sign and ordered all copies of its report torched. This was the seed of the public's deep-rooted suspicion that the government would never come clean on the true nature of UFOs.

The immediate consequence of Sign's demise was a new investigation, Project Grudge (the name bespeaks the

**Remember, this was before elaborate radar systems had been constructed to detect intruding aircraft. NORAD, the North American Aerospace Defense Command, was initiated in 1958.*

motivation). This effort was soon supplanted by Project Blue Book, an attempt to establish an aura of unimpeachable honesty for its examination of the mysterious lights in the sky. (The moniker alludes to the pledge of honor that college students, who are obviously untrustworthy, have to write and sign in their exam booklets.) This project lasted from 1952 until 1969. Other UFO investigations included one at the University of Colorado and a study by an assortment of distinguished academics known as the Robertson Panel. (Among its members was physicist Lloyd Berkner, the onetime director of the National Radio Astronomy Observatory who had prompted Frank Drake to cook up Project Ozma. Also on board was physicist Luis Alvarez, who later showed that the dinosaurs were most likely annihilated by a rock from space.)

In other words, for two dozen years UFOs were subject to more scrutiny than Lindsay Lohan's social calendar. The conclusion of all these panels and projects was pretty much what you wanted it to be. You could read them either way: The aliens were here, or they weren't. These studies didn't resolve the UFO issue; instead, they mostly hardened attitudes.

To some extent, this regrettable result was inevitable, implicit in the very nature of the exercise: a broad review of a large, and highly heterogeneous, database. Some of the reports were well documented, with multiple witnesses and photos or films. Others consisted of no more than the testimony of one or two individuals. To absolutely rule out the premise of alien visitation, the investigators would have to find a nonextraterrestrial explanation for every one of these cases. There was no way that could happen, and it didn't. Blue Book examined roughly 12,000 sightings. It could explain most of these as aircraft, birds, bright planets, and a laundry list of other prosaic phenomena. But Blue Book failed to find plausible, non-alien explanation for about 700 of the reports. This percentage was typical: Most UFO investigations have ended up with a residual of unexplained cases that tally somewhere between 5 and 10 percent of the total.

In 2008, the British government released—to considerable fanfare and salivating expectation—the first of a large number of Ministry of Defence UFO reports dating back to 1981. Once again, most of the sighted phenomena had prosaic explanations, but between 5 and 10 percent were deemed mysterious.

The unexplained cases are tantalizing to some people, since it encourages them to opine, "Sure, most of these reports are bogus, but a few are actual alien spacecraft." That's possible, of course. But you can't draw such a conclusion from these studies, unless you're perversely allergic to logic. According to the Department of Justice, 39 percent of all the murders in the United States during 2006 were unsolved. You might hypothesize that some of the victims in these still open cases were snuffed by aliens. But it's doubtful you'd make such a claim in educated company. Just because a UFO sighting is unexplained doesn't mean that the explanation is unusual.

I'll also note an important asymmetry in logic here. While disproving the alien hypothesis demands getting to the bottom of every one of the thousands of UFO reports, proving it requires no more than showing that *one* case is nonterrestrial. Accomplishing the former is, in the real world, insurmountably hard. Doing the latter—if Earth is really being buzzed by uninvited guests—should be easy. But the fact that this hasn't happened, either to the satisfaction of the half dozen investigative panels, or at a level that convinces many scientists, is highly significant.

UFO converts point to the enormous number of reported sightings as if the tally should convince you that aliens really are afoot. In fact, it shows the opposite. If, among all those incidents, not one has produced unambiguous evidence, not one is good enough to display in the Smithsonian, then the alien hypothesis confronts a monumental failure. An abundance of equivocal sightings fails to add up to a compelling case for visitors.

WHY I'M SKEPTICAL

By now the attentive reader may sense that I doubt that our planet is actually host to interstellar visitors. Is this squinty-eyed disbelief merely a requisite of my career? Maybe I don't like the idea that the aliens are wandering the globe because I'm part of a project attempting to find them elsewhere, at light-years' distance.

On the contrary, nothing could be more exciting and profoundly important than to know that Earth has shown up in some aliens' telescope and prompted them to drop by. Frankly, if the evidence were good enough, my colleagues and I would abandon our antennas and begin crawling the countryside. It would be easier and cheaper. It would also offer the tantalizing possibility of communication that was up close and personal.

But after more than 60 years of UFO sightings, we still seem unable to come up with the good stuff. Physical evidence—a taillight or knob from an alien craft—is in short supply. Suppose you were given the task of proving that ships from Asia were visiting California. A stroll to the nearest container port would clear that up rather quickly, but even just walking the beaches would soon provide you with plenty of physical evidence: bits of cargo (remember all those Nike shoes?) and other detritus that have washed ashore. Somehow, alien spacecraft never leave such a hardware trail behind.

Flying saucers are Exhibit A for those making the case that aliens are among us. But they're not the only evidence. Among the other bizarre phenomena (why are they always bizarre?) blamed on our cosmic confreres are crop circles, abductions, and architectural mementos left on Mars and the moon.

The first of these, crop circles, first started sprouting up in the early 1970s. Since then approximately 10,000 of these agrarian greeting cards have appeared, providing puzzlement to the public and inconvenience to farmers. Years of investigation have established that these grain graphics are easily constructed by humans (students seem particularly adept and motivated), although a die-hard core of believers maintain that a few are the product of astral artists. The claims of alien authorship are flimsy, based on the way the wheat stalks are bent, or the fact that some of the patterns are ambitiously complex.

These plant patterns seem oddly inconsistent with true alien messages, however. Even aside from the fact that they usually materialize overnight or at the start of school holidays—even aside from the curious fact that most of these graphics have appeared in but a few counties in England (have the aliens no interest in speaking to anyone else?)—they simply don't make sense as a method of interstellar communication.

This is principally because wheat fields are poor memory storage devices. Even the largest crop circles have, at most, only a few hundred to a few thousand bits of information. I'm speaking of "information" in a mathematical sense, which is a measure of the maximum amount of actual "cultural" information these patterns could convey. As a trivial example, the simplest form of crop circle—just a circle—can be described by a single number: the radius. Two circles have a few more numbers (radii for both, and the offset position of the second circle with respect to the first), and so on. A famous glyph that appeared near the Chilbolton radio observatory in Hampshire, England, in August 2001 mimicked a deliberate radio broadcast to the stars made nearly three decades earlier from the Arecibo Observatory. This was one of the most complex grain graphics ever. Even so, its information content was a paltry 1,679 bits. The amount of information in this paragraph exceeds that.

Indeed, writing on a sheet of paper conveys data far better than chopping crops. Given the prodigious difficulties of interstellar travel, you probably wouldn't expect visitors to communicate with smoke signals once they got here, yet crop circles are similarly limited, ambiguous, and ephemeral. I'll also point out the inconvenience that the fields are open for communication purposes only two months of the year, when the wheat's in residence. Additionally, farmers mow down these uninvited defacings almost immediately to prevent tourists from mashing down the rest of their harvest.*

If you still believe that aliens would travel hundreds of light-years to carve temporary graffiti in our wheat, then your imagination is one of the seven wonders of the world, and should be bronzed.

**Occasionally, the farmers reckon that it's more profitable to open their suddenly famous fields to the public and charge admission rather than to try to make a living with agriculture. In such cases, the glyphs are temporarily left standing.*

Abduction by aliens, with its attendant and frequently impertinent probing, is a phenomenon as well known as the crop circles. And, unlike the latter, abductions don't require you to live in southern England to be directly involved. Indeed, the aliens seem to be shanghaiing people everywhere. The template for many alien abductions was laid out in *Communion,* the 1987 best seller by horror fiction writer Whitley Strieber. In this book, Strieber describes humanoid creatures who are poster children for today's thin-lipped, big-eyed aliens. The "visitors" (Strieber doesn't explicitly say they're extraterrestrials) broke into a cabin in which he and his family were sleeping in upstate New York during the winter of 1985. They had pale complexions and wide faces. During Strieber's interaction with the intruders, they arranged to probe both his brain and his rectum.

About a half dozen years ago, I participated in a panel discussion on extraterrestrial life that was brought to a standstill when an older gentleman rose to his feet and declared the following:

"I'm not quite sure why you're telling us about these efforts to find radio signals from the sky. The aliens are here on Earth—many species of them—and I have been routinely abducted ever since the age of six."

This brave admission was followed by more:

"I would like everyone in this room who has also been abducted to rise."

The audience was overfull, with people standing around the edges. The upright attendees looked at one another and quickly sat down on the floor. I expected that the self-declared abductee would be the only one left on his feet. I was wrong. Seven other spectators slowly stood.

I made a quick mental calculation. If this audience was typical of the world population, then roughly 30,000 people are being abducted *every day*. That's major. That's a hundred times the death rate from automobiles and war put together. Someone would notice.

Indeed, abduction stories *have* been noticed—and investigated by psychologists. A phenomenon known as sleep paralysis may be causing the experience of those who believe they've been hauled off by little gray guys. Sleep paralysis is a fairly common occurrence for people who are passing through that delicious transition between sleep and wakefulness. They're conscious but still have trouble moving their bodies. Many folks also have the impression that someone is standing next to their bed when they're in this state. So sleep paralysis can account for many of the abduction stories. Of course, aircraft can account for many of the UFO stories too, but that neither assuages those who suspect an alien connection nor disproves their claims.

Nonetheless, and as with the crop circles, there's a strong scent of something odd wafting from the abduction idea. Why are visitors to our world only interested in freaky behavior? Abducting single individuals for personal inspection is not only inappropriate, but suspiciously convenient. They're here, but we have only witness testimony as proof. Given the enormous number of people who claim to be abducted over and over, why has no one been prescient enough to take along a camera?

Nonetheless, virtually none of the daily drip of e-mails and phone calls that I get from people who believe they've experienced encounters with aliens smack of chicanery. Those seeking to get in touch about strange craft or strange incidents seem genuine, even if they've failed to offer even one event that I thought sounded possibly extraterrestrial in origin. They're not lying to me.

But there's another category of evidence that screams for skepticism: the claims of deliberate alien constructions on the moon or Mars.

MONUMENTS ON MARS

The routine postings by NASA and other space agencies of high-resolution photos of lunar and Martian landscapes, available to all on the Internet, has democratized image analysis of other worlds. No longer the exclusive provenance of rocket scientists and planetary geologists, pictures of the planets and moons of the solar system can now be examined by anyone with the interest and the time. To say that this data set is rich is to understate its wealth; photos from the HIRISE telescopic camera on NASA's Mars Reconnaissance Orbiter can capture red planet detail that's the size of a large bunny.

Ergo, it's hardly surprising that I occasionally receive missives from amateur sleuths who've found strange figures in this imagery. One gentleman sent me a Mars photo in which he recognized the silhouettes of more animals than populate the zodiac, all of which he took to be conclusive evidence of extraterrestrial activity (although it's unclear why Mars would be decorated with *our* fauna).

Our brain is wired to see and recognize faces, both human and animal. Even Leonardo da Vinci noted that if you stare at the clouds for a minute or two, you'll easily pick out faces. Indeed, the talent for recognizing faces runs so deep that people espy the Virgin Mary in tree branches, bridge supports, pizzas, and most famously, a decade-old cheese sandwich. This penchant for recognizing faces is part of a phenomenon called pareidolia, and we all have it.

Pareidolia has led to a dismaying sideshow in the hunt for extraterrestrials. As noted, space research of the past half century has quashed our hopes for intelligent beings on Mars. Yes, life on the red planet could exist, but we've downgraded our expectations from the telepathic beings of *The Martian Chronicles* to senseless, single-celled pond scum.* Clearly, for those raised on sci-fi tales of elaborate Barsoomian societies, this is a first-rate disappointment. Consequently, some people on the fringes of research take advantage of the wish to bring back the Martians. If the academics won't reinstate the hydraulic engineers of yore or something at least as good, these peddlers of fanciful science are keen to try. And pareidolia plays a pivotal role.

Consider Richard C. Hoagland, a frequent fixture at UFO-related events and a garrulous guest on late-night talk radio. Hoagland, who was once a curator at the Springfield Museum of Science in Massachusetts and who also worked briefly at NASA's Ames Research Center, has built much of his career around an early public relations gaffe by the space agency. In midsummer of 1976, while the Viking 1 lander was pawing the Martian dirt at Chryse Planitia, its orbiter was making reconnaissance photos from nearly 1,200 miles (1,900 kilometers) up, trying to find a suitable touchdown spot for Viking 2. In the growing stack of pictures compiled by the orbiter was an aerial shot of a mesa- and mountain-strewn region known as Cydonia. Among the gnarly plutons hunkered down on the Cydonian plains was a rock that resembled a mile-wide face. It occurred to the NASA public relations people that they might gin up public interest in the

---

**There may not be any ponds, however. Rather, these possible red planet residents would live in subterranean aquifers.*

Mars program by whimsically pointing this out. They released the photo, remarking in the caption that "the huge rock formation in the center, which resembles a human head, is formed by shadows giving the illusion of eyes, nose, and mouth."

Innocent enough. But some people, and especially Hoagland, took NASA's casual comment to places it was never meant to go. They suggested that the mesa photographed by Viking 1 wasn't just a geological blemish turned into a face by trick of light and shadow and the penchant of our brains to see faces everywhere, but a deliberate construction. It was a face built by aliens. Hoagland soon elaborated this story. He pointed out odd-looking features within a few dozen miles of the head which he claimed were part of a "city," nicely fitted out with a pyramid, a fortress, and other clearly essential municipal constructions. The Face on Mars was just an outlying structure in an ancient Martian metropolis.

Incredulous, NASA pooh-poohed the whole thing, but Hoagland claimed that the space agency was being disingenuous, and some among the public agreed.

In 1998, a new NASA spacecraft—the Mars Global Surveyor—took up its orbital perch around Mars, its cameras fixed on the terrain below. This was the first craft able to make high-resolution photos of the red planet since the Viking orbiters nearly two dozen years earlier. But it was far superior to its predecessors: Its cameras could resolve details ten times smaller than the Vikings could. Hoagland and his followers pressed NASA to make a new Face shot, but the agency, in a studied lack of sensitivity to the public's more fervent beliefs, initially refused to spend the time and thruster fuel.

From the agency's point of view, this showed an appropriate disdain for the gonzo idea that a stone head had

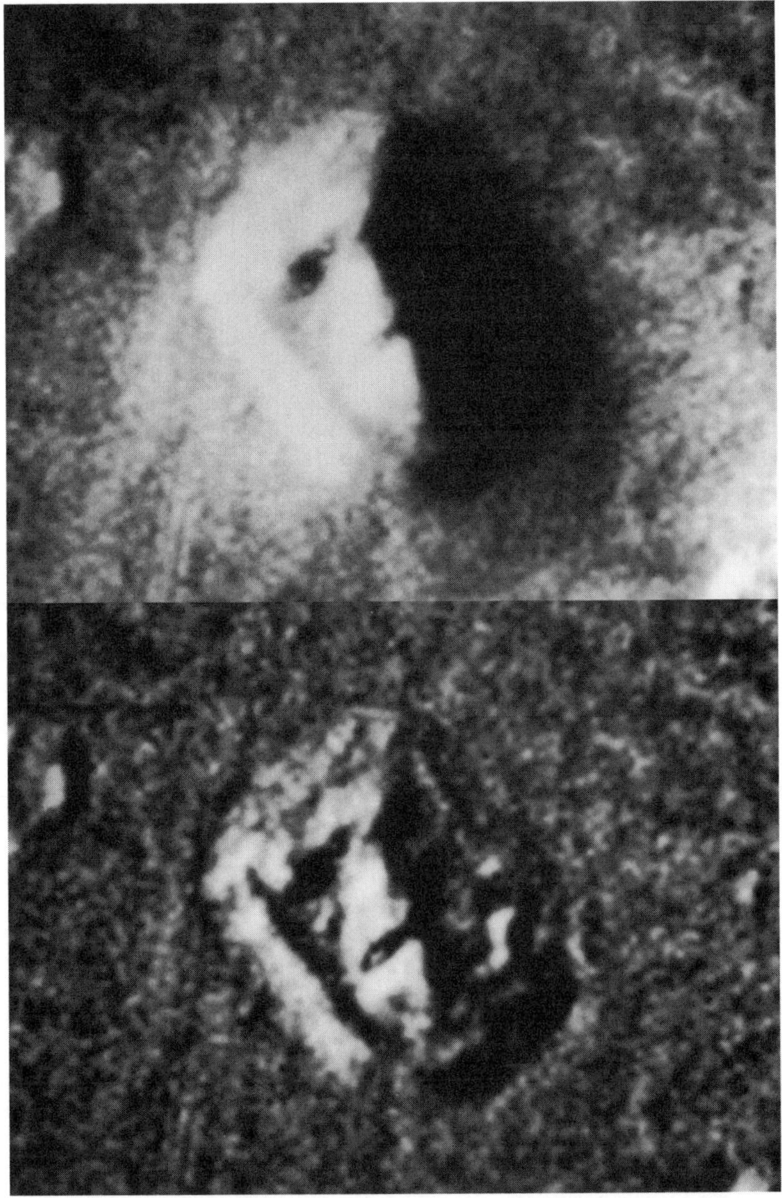

In 1998, high-resolution photography revealed the "Face on Mars" (top) to be merely an interesting-looking mountain (bottom).

been built by bored aliens visiting Mars. Of course, it also served to ratchet up the claims of conspiracy theorists, who could now chant with greater volume that NASA was engaging in secrecy and subversion. Finally, someone at the agency recognized that *not* photographing the Face might be more costly (in terms of public support, which eventually translates into congressional funding) than imaging what was, as far as the scientists were concerned, a random and largely uninteresting tract of terrain. So NASA announced that on April 8, 2001, Surveyor would make a new and detailed picture of what had become (in the public's mind, at least) the most famous 300 acres on the red planet.

The evening before Surveyor snapped the momentous photo, I was called by Glennda Chui, a science reporter for the *San Jose Mercury News*. Glennda was interested to hear my impression of the Face, and whether I expected major revelations from the new photos. I suggested that she could do an equivalent experiment at home, by driving to the nearest supermarket and purchasing a ten-pound bag of potatoes.

"I'm pretty sure," I said, "that if you look at each of those potatoes, you'll find a face. But I don't think it will occur to you that this might be some effort by the spuds to get in touch." Call me conventional, but I figured the Face was just a mountain.

When the higher resolution photos were released, they showed . . . a mountain. That is, they showed what looked like a mountain to most people, including planetary geologists. NASA, once again trying to appear bemused, quickly plotted the best route for future mountain climbers to assault the Face and put a chart of this path on the Web. It turns out not to be a particularly challenging ascent, incidentally—except, of course, for the lack of much air.

Richard Hoagland, who had written a book on this Martian mug's artificial origin, was unfazed. The high-resolution pictures, he claimed, merely confirmed what he had been saying all along. Sure, the details were not particularly face-like, but what do you expect on a planet where windborne dust whipsaws the rock, day after day? It was a *ruined* face, that's all. It may not surprise you to learn that Hoagland's supporters were convinced by this backhanded rationale. The Face's fame lived on.

In 2006, the Europeans took their best shot at the Face, using the Mars Express orbiter to make, well, the best shot ever of this topographic zit. It showed, once again, a mountain, although in greater detail than earlier snaps. Sure, this new evidence was like kicking someone who was already down, but Hoagland wouldn't give up. He kept whistling the same weird tune, and when the Spirit and Opportunity landers made their first ground-level photos of the Martian landscape in 2004, Hoagland soon declaimed on the airwaves that various pieces of alien technology could clearly be seen in the pictures, stuck in the red sand like detritus in a half-buried junk yard. He claimed that NASA was, yet again, ignoring the evidence.

These days Hoagland has relegated the Face story to the back burner. A new, wonderfully detailed photo made in early 2007 by the HIRISE camera has shown yet again that it's a mangled hunk of rock. If this is a face, it's one that's been soaked in battery acid. So Hoagland doesn't talk as much about it as before, although he still has a following for his "evil NASA" theories. This is despite the obvious fact that nothing, *nothing* would guarantee a major boost to the agency's funding as fast as finding even the slightest hint of intelligent beings on the red planet. It's the very last thing they would ever cover up, even if they were inclined to cover-ups. Somehow, this point escapes most of Hoagland's devotees.

It might occur to you that the Hoagland story and the earlier protestations of Percival Lowell seem similar. Both claimed to see evidence of intelligence writ large on the sands of Mars. But there are major differences. Lowell was a professional astronomer, with academic credentials (yes, that's right—as bad as it may sound, he actually had extensive schooling in science and math), and the ability to build his own observatory—a facility that he used for decades to investigate various puzzlements of the solar system. Hoagland simply looked at NASA imagery (which he occasionally "reprocessed") and said he saw things that others didn't.

DO VISITS MAKE SENSE?

Some people have tried to decide if UFOs are extraterrestrial hardware by mulling over not the sightings themselves but the external factors: Could alien rockets even come here? Is there evidence they visited ancient societies? Have our nuclear weapons tests encouraged them to travel to Earth to save us? And so forth.

First off, let's admit that interstellar travel doesn't violate the laws of physics. After all, the Pioneer and Voyager probes, launched in the 1970s to reconnoiter Jupiter and Saturn, are three decades into an inadvertent interstellar journey. They passed Pluto long ago, and are headed out into the galaxy. The kicker, of course, is that, despite the fact that these craft could traverse the U.S. in six minutes, they'd take many tens of thousands of years to cover the distance to even the very nearest stars.*

In other words, while traveling between the stars is not impossible, it may not be practical. Consider that the closest extraterrestrial society to us is probably hundreds of light-years distant, if not more. If alien rockets are no better than our own, then their travel time to arrive at Roswell (or some similarly glamorous destination) is millions of years. That's a long time to be squirming in a coach seat. The aliens are likely to book passage for such tedious travel only if they live for millions of years or more.

Consequently, the following conclusion seems unassailable: If extraterrestrials have chosen to visit Earth, they're blessed with rockets far speedier than ours. Indeed, to travel between galactic habitats in less than a millennium, their spacecraft must travel at a substantial fraction of the speed of light.

Is this possible? Many earthly scientists have considered how such a rocket might work. Chemical rockets are, as you might surmise, entirely inadequate for such velocities. Nuclear fission or fusion rockets seem marginal as well. But a matter-antimatter rocket could achieve speeds of tens of thousands of miles per second (indeed, that's how the starship *Enterprise* is powered, according to *Star Trek* lore). That leaves aside the difficulty of making and storing the antimatter, of course, but that's a problem left to the student.

**These spacecraft are not, of course, aimed in the direction of any particular stellar targets, and won't even get close to a star for a very long time. For example, after 40,000 years Voyager 1 will pass within 1.6 light-years of a faint star known as AC+79 3888 in the constellation Camelopardalis. Its sibling, Voyager 2, will slide within 4.3 light-years of the bright star Sirius in the year 300,000.*

Another approach to truly speedy travel is to forget about bringing your own fuel, and simply collect it on the way, somewhat akin to the way horses operate. The physicist Robert Bussard has proposed an interstellar rocket that scoops up interstellar hydrogen gas in front of the craft. The gas is compressed and fused into hydrogen (just like an H-bomb), with the sped-up atoms shot out the back. Since such an engine can operate continuously, an interstellar ramjet might reach extremely high velocities. And, like a sailing ship, it could go anywhere, since there's always more fuel available just ahead of the rocket.

Alas, this design has problems. To begin with, the fuel is highly diffuse. The wispy gas between the stars is breathlessly thin, typically amounting to only 10 to 100 atoms per cubic centimeter. In the laboratory, this density is referred to as an "ultrahigh vacuum" not a "cloud of gas." So the ramjet needs a nose scoop wider than the Earth to corral its fuel. It's also unclear how to make the fusion process work. Still, it's conceivable that advanced aliens, somewhere in the galaxy, solved these daunting problems long ago.

Mind you, just having a fast craft isn't all that's required. Interstellar material, both gas and dust, would be hitting the front of the passenger compartment at enormous speeds. This would produce an endless torrent of hard radiation that could easily kill both the onboard electronics and the onboard crew. The aliens had best be wearing lead underwear several feet thick. And encountering anything as large as a sand grain could be fatal. At half the speed of light, such a grain would dent the windshield with the energy of 10,000 sticks of dynamite. That's no ordinary bug spot.

The formidable problems of rocketing between the stars hasn't been lost on the mass of folks who think alien-operated

UFOs are dropping into our skies. Some of these people suggest that the manifold obstacles that beset such craft will pleasantly evaporate in the heat of new physics. In particular, we should assume that the extraterrestrials have mastered wormholes or warp drive, either of which would allow them to reach their destination faster than a light beam, and do so by bending space.

Numerous and highly intriguing schemes have been proposed to do this. Alas, most of them require marshaling massive amounts of energy or rounding up exotic material that might not even exist. Scientists will not say that such schemes are impossible. We also can't say whether they *are* possible, because theories in this field are still incomplete. Yet even if one of these schemes eventually looks right on the blackboard, there is no guarantee that it is feasible in practice.

The essential problem is this: The distances between the stars are sufficiently daunting that piloting a rocket to Earth seems grossly impractical. Indeed, some experts opine that it's flat-out impossible. Consequently, to explain the alleged fact that we're being visited, many UFO proponents resort to physics (and technologies) that are still hypothetical. This is an argument from ignorance. "We don't know how they could get here with any practical scheme, so we're suggesting that they've mastered warp drive. Which is something else we don't know." It's imaginative. It's not convincing.

A second discussion point concerns the odd fact that the aliens have decided to visit now. Earth has been paved with life for roughly four billion years, and they've decided to visit only at the end of the 20th century? Why is that?

One explanation, generously offered on blogs or call-in talk shows, is that the aliens feel obliged to come here to correct our bad behavior. They've seen the bright flashes of our nuclear bombs, and have reluctantly chosen to intervene before we either exterminate ourselves or become a threat to others. A slightly different version of this "alien police" hypothesis is that extraterrestrials have noticed our environmental depredations and are juiced about the need to save Earth's ecosystem.

Well, it's hardly clear that the aliens would have the slightest interest in promoting our peaceful coexistence, reducing our greenhouse gases, or plumbing our reproductive systems. I've noticed that I don't really get too worked up about the vicious wars of the ants in my backyard, let alone feel the need to probe their tiny privates.*

However, whether or not they'd be motivated to help us or handle us doesn't answer my question: Why are they here now? That question has force because there's no way the aliens could be aware of our existence.

Homo sapiens is desperately hard to find when you're light-years away. It's often said that astronauts can see the Great Wall of China from orbit, and that's true. You can also see lots of other stuff, including highways, airports, dams, and even jet contrails. But from the moon, none of these things is visible without a sizable telescope. From Mars, all earthly activity becomes *invisible* even if the Martians are outfitted with the equivalent of our *best* telescopes. Alpha Centauri, the nearest other star system, is 750,000 times farther than the red planet. Earth shrinks to a dim, barely luminous point

---

**Mind you, if the ants invade my kitchen, my disinterest in their behavior changes quickly, and I attack them with toxics and a vacuum cleaner. Given the reputed loutish behavior of some of the aliens said to be invading Earth, it's surprising how little we've fought back. Probing is apparently no more than an inconvenience for the populace.*

at such remove, and all the works of mankind are no more than very fine fuzz on an incredibly distant peach.

Indeed, from the stars, the only manifestly detectable signs of humanity's existence are our high-frequency, high-power radio, TV, and radar transmissions (forget the nuclear explosions, by the way—they're not that bright, and they don't last very long). These detectable broadcasts go back about 70 years, roughly to 1940, which means the first of them are now rippling through the cosmos 70 light-years from where we sit. But the aliens are said to have arrived on Earth no later than 1947, when they buzzed the Cascades and then pitched into pastureland near Roswell—only seven years after we started making our presence known.

This chronology compels us to consider the following scenario: Suppose aliens picked up our 1940 broadcasts on their radios, and made an instantaneous decision to save our world. To do so, they hopped into their spacecraft and set a course for Earth. Let's further assume they have wormhole-capable rockets and they arrived in only seconds or less. Consequently, to reach our skies by 1947, their home planet must be seven light-years away or less. That's next door. In fact, there are only two stellar regimes within that distance: the triple-star Alpha Centauri system, and Barnard's star. What are the chances that one of these two nearest star systems houses intelligent beings? Bluntly stated, they're stupefyingly small.

If aliens have come here, it's not because they know we're here. They don't.

But maybe we're making too much of this illogical premise. Maybe they happened to arrive during your lifetime by sheer coincidence. That could be, but the probability that a random arrival of aliens, one that could occur at any point in the 4.6-billion-year history of Earth, will by chance coincide with your existence on this planet is about 1 in 50 million. That's a long shot indeed, about the same as winning the California Super Lotto jackpot with your next ticket.

Perhaps another claim should be considered. They're here now simply because they're *always* here. Or perhaps they're routine visitors. After all, if the aliens visit regularly, their presence since 1947 wouldn't be especially remarkable. Indeed, some folks would be happy to endorse the idea of frequent alien sojourns to Earth, because it supports their contention that extraterrestrials have helped some (apparently inept) societies with their public works projects. For example, it's often suggested that otherworldly types were paying a call 5,000 years ago to teach the pharaonic Egyptians how to stack blocks of limestone into giant pyramid shapes. Others allege that alien engineers were consultants to the Nasca Indians, who were busily decorating the Peruvian desert floor with glyphs of turkeys and other of their favorite fauna about a thousand years back. (I note that, for some disturbing reason, no one suggests that aliens may have assisted in building the Parthenon or Colosseum.)

If any of this is true, it argues for either continuous extraterrestrial presence or visits every few millennia. The former seems unlikely because—if they've always been here, why haven't they taken over the Earth? Or at least modified it in some significant way? They haven't left any recognizable trash, let alone traces in the fossil record. The repeated visit scenario is also unlikely, because it would imply that there have been *millions* of alien expeditions to our planet without any result. Which alien society would fund that?

When we step back and consider what's really being hypothesized about visiting aliens, we see it's all about us.

They're here because they want to help, or at least engage us in inappropriate behavior. It's like a cosmic makeover show: Someday, someone powerful will improve our lives.

We are clearly of great importance to the aliens. Sure, they haven't actually solved any of the global threats that face us, and true, their personal attentions don't ever seem to lead to durable relationships. But at least they're interested, something your human neighbors might not be.

However disappointing it may be, neither logic nor the numbers encourage believing in this kind of egocentric encounter.

HOT-BUTTON ISSUE

I've expressed my skeptical take on the UFO issue to the media, during talks, and in dozens of articles. As a consequence, it takes nerve to open my e-mail. You might think that correspondents who are convinced that Earth is being visited would simply tell me why they think so. That certainly occurs. But what also occurs—and frequently—is warfare. Allow me to treat you to a very small sampling of this corrosive commentary, excerpted from several e-mails:

"You're the most closed-minded person I ever heard on the radio. You should step down and shut up."

"I find your group and you to be really stupid! I would hope that someone there can do research into your own group's history and find information on meetings that took place in 1977 with SETI explaining fully that the aliens are visiting this planet."

"I think you should retire and let someone with guts take over the project and look right here in our backyard for ET."

"They are here, I have seen them on two occasions quite unmistakably . . . you are all about the money, and being able

to sit on your fat asses doing as you please with our money. I have no respect for the likes of you at all."

"I understand from my link to God that there are nice people and evil ones. On Earth, you are evil."

You might think that, since both the UFO fans and SETI are intensely interested in the possibility of intelligence on or from other worlds, they would treat one another with mutual sympathy. Or at least some understanding. That doesn't seem to be the case, or at least, it's often *not* the case. I frequently get correspondence from people who have witnessed something mysterious and want my opinion on what it might be. That's a straightforward interrogatory, and I try my best to make suggestions. Less often, I receive missives from people who claim personal involvement with aliens and simply want to describe their experience. I'm sympathetic. But the majority of correspondence is accusatory and highly critical. These correspondents don't want to talk about sightings or abductions. Rather, their purpose is to excoriate me for not admitting that the aliens are here. It's painful to realize that the matter of visitation is deeply, viscerally emotional, with both believers and disbelievers frequently resorting to nasty argument and ad hominem attack.

Why might this be? Well, let's first consider this from the skeptics' point of view. Scientists are often inclined to ignore the UFO crowd because, as Marcello Truzzi wrote, "extraordinary claims require extraordinary proof" (this aphorism was later popularized by Carl Sagan). If UFO supporters are going to state that they've found sentient life from the stars, they'd better offer more than eyewitness testimony as proof. Just because a wide smattering of people think they know what they saw doesn't mean they do. After all, you could find millennia of eyewitness testimony that the Sun orbits the Earth.

In addition, the occurrence of occasional hoaxes has inclined many academics to label all extraterrestrial claims about UFOs as pseudoscience. They regard UFOs to be a phenomenon similar to remote viewing or talking with the dead. Indeed, the cynical would argue that the scientists who do SETI *have* to feel this way, for otherwise it's hard to understand why they're spending their time hunched over a lab bench, building digital receivers, when they could be checking out UFO reports.

There's also the matter of publicity. Flying saucers, abductions, and the like get a lot more air time than nerdy scientists, and academics may resent this. I recall the first TV program for which I was interviewed at the SETI Institute, only months after my arrival. I was chary about standing before the camera, being the new kid on the block, but the producer assured me that this was going to be a "responsible show." She sent her completed (and very responsible) program to the network, only to have it quickly returned bearing a yellow sticker demanding "more abductions." Science was deemed not interesting enough, so bring on the prod-prone grays.

This parochial focus does make sense. Compare the stories: The UFO believers have not only found the extraterrestrials, they've found them nearby. It's easy to see that this narrative trumps one where academics maunder on about a possible future detection of some radio static.

The very practical matter of competing for support also plays a role. SETI scientists have a long-standing fear of being tarred with something called the "giggle factor." They worry that their explorations will be caricatured as amusing rather than profound, as activities akin to scouring the woods for leprechauns or the tooth fairy. So SETI feels a need to distance themselves from UFOs, simply to forestall conflation of the two activities by potential SETI supporters. If you're trying to raise research dollars from the government or serious donors, credibility counts.

OK, so the academics are annoyed by the UFO crowd. But why are the latter so often enraged at the former?

Antenna envy is surely one reason. The scientists have impressive high-tech equipment, not to mention professional affiliations. They, and their lab gear, look like serious stuff on camera. So despite the fact that the UFO believers claim to have succeeded in actually finding extraterrestrial intelligence, they resent not getting the requisite kudos. And indeed, establishment science really doesn't accord them any respect: Few academics spend a minute of their time investigating the UFO phenomenon.

UFO fans regard themselves as heroes, denied a rightful place in the pantheon of discovery (and, one presumes, their Nobel Prizes as well). Besides being heady stuff, it's empowering to believe you know something of blazing importance that those ivory-tower types don't, despite their fancy titles and tweed jackets. It's class warfare: the proletariat versus the privileged.

Finally, for UFO adherents lurks the mother of all worries: SETI might succeed. How much attention would the public lavish on stories of lights in the sky or gray guys in the bedroom if researchers around the globe could find and intensively study a signal coming from another world? Keep in mind that UFO sightings are neither predictable nor, in large measure, repeatable. Something is seen, and maybe photographed. But a signal from space is accessible to anyone with the right telescope (and this amounts to a large and international group of people). You could point at the source of the signal, tell what star system it was coming from, measure its distance, and maybe even understand it. You'd have hard numbers, and surely some new knowledge. Collected data could be distributed on the Internet, with everyone taking a shot at comprehending what they meant. Unlike putative saucer debris or alien bodies, this signal source couldn't be spirited away by paranoid men in black and stacked up in a secret Nevada warehouse. The discovery would be exposed to all, up in the sky. With evidence like this, UFOs would have a hard time staying in the limelight.

For those invested in the idea that the aliens are here, the specter of a SETI success is downright scary.

CONVINCING EVIDENCE

Could anything wean me from my skeptical take on UFOs? Something that would tempt me to think that we're really being visited?

Yes, of course.

But convincing me won't be a trivial exercise. The major reason I'm reluctant to believe is the dicey nature of the evidence. If Earth is really hosting extraterrestrials, why is this so difficult to prove? If you had asked the Native Americans 60 years after Columbus's first foray to their continent if they believed alien visitors were among them, they wouldn't have hesitated to say yes. Indeed, by 1550 many natives were slaves to the Spaniards, and enormous numbers of them had died of disease. In Central and South America, mighty Indian empires had crumbled.

The locals knew they had visitors (who soon became permanent residents), and they didn't have to hold heated debates about it. And yet 60 years after the first modern flurry of UFO sightings, the presence of aliens is still a doubtful claim.

I've repeatedly asked members of the UFO fraternity to explain the reason for this continued ambiguity. Where's the physical evidence? Why can't I find exhibitions on alien visitation at my local science museum—they would surely be popular. In addition—and this is highly significant—why aren't tens of thousand of academics examining this phenomenon? For instance, if biologists, zoologists, and primatologists thought that there was the slightest chance that bigfoot really existed, they would be stomping the Pacific Northwest into fresh mulch looking for him.

The response to my queries almost invariably boils down to two excuses. First, the UFOers allege that the academics' skepticism stems from their own pigheaded reluctance to look at the reports. They've already made up their minds.

The second excuse is that good physical evidence does exist. It's just been hidden away by a government anxious to keep the populace from rioting in the streets.

Well, this isn't very helpful. The first argument, that scientists are close-minded, blames the buyer for disinterest in a shoddy product. Yes, many scientists will ignore evidence of a phenomenon or explanation that's new, one that contradicts conventional wisdom. But not all do. Indeed, in a matter as potentially dramatic as visitors from elsewhere, the general apathy of researchers is hard to fathom. Science exalts open-mindedness, even though it isn't always practiced that way. If the evidence were good, you'd have plenty of prestigious investigators.

The second contention, that the really convincing artifacts have been scooped up and secreted away by the government is, once more, an argument without legs ("there's good proof, but I just don't have it"). More than that, it doesn't pass the smell test. Since UFOs are seen worldwide, this explanation requires that *all* governments—not just the presumably malevolent Americans—cooperate to hide the truth.

So what would it take to swing my own opinion around? To begin with, I'd like better images. After all, several sightings are reported every day, so surely someone can take good pictures or videos. In general, the photos and films that show the most detail are those that are clearly hoaxed. Imagine if you were asked to prove the existence of blimps. If you gathered photos from the last, say, 60 years, you'd find some good ones. In fact, you'd find a *lot* of good ones. The images wouldn't be, as the UFO photos often are, inconclusive, low-detail pictures—images in which nearly every craft looks different. In an era in which point-and-shoot cameras fit in a shirt pocket, and cell phones can take photos, why is this alien aerial presence so difficult to record?

Second, I would be pleased if the UFO crowd would bring on some decent physical evidence. Something better than burn marks in the grass or amorphous bits of rubble pried out of an abductee's epidermis. Stories are entertaining, but scientists need evidence they can test in the lab.

Third, why have these craft not been spotted by one of the many satellites that pirouette above our planet? Everyone watches weather satellite video loops on the nightly news. Since the high-tech hardware that makes this imagery is designed to monitor storms churning across the landscape, the satellites are fixed in the sky—in geosynchronous orbits far above Earth's surface. However, their cameras can only see detail half a mile in size or larger. They could arguably miss alien craft. But of course we have an abundance of better imagery around. NASA's new Terra satellite for monitoring the environment has a wide-spectrum camera that can spy objects as small as a bus. The French company Spot Image is happy to sell you (and has sold to Google Earth) satellite photos of Europe that will reveal anything half that size. In the United States, GlobeXplorer's AirPhotoUSA will provide aerial photography that can pick out objects comparable in dimensions to a coffee-table book.

Now it has to be said that the imagery that's being taken all the time, and therefore would likely include any untoward flying craft, is generally of low resolution. High-resolution photos are either made infrequently (for any given area) or are restricted to certain geographical regions (e.g., Europe or the U.S.). Of course, the military (of several nations) surely has better reconnaissance coverage than the commercial satellites, but if the UFO crowd is to be believed, the military is the problem, not the solution. Nonetheless, every year new satellites—from a multitude of countries—are heaved into the sky, and anyone who has looked for their house on Google Earth knows that the publicly available imagery is getting better all the time. You could still make the case that all these satellites have failed to find alien UFOs. But reconnaissance capabilities are now in the hands of many nations. To claim that orbiting "birds" have actually seen extraterrestrial craft but their owners are not telling us requires not just an American conspiracy, but a global one. In any case, each year that goes by sees an increase in widely accessible imagery. Frankly, I would switch my point of view on UFOs in a New York minute if I saw a satellite photo of an alien craft that's as crisp as the Google Earth imagery showing the Honda parked in my driveway.

The curious failure to find UFOs extends to the scrutiny of the sky from the ground. Probably the best known facility engaged in this sort of work is at Cheyenne Mountain Operations Center in Colorado. The radar systems there are all-weather, and altogether impressive in their ability to find and track thousands of objects. Since they can see something the size of a softball in orbit, they could also see saucers that are considerably closer to the ground. Why don't they?

Once again, given that this is an operation run by the military, some people will allege a cover-up, and sci-fi fans will suggest that the alien craft have cloaking devices (although these don't seem to work well enough to prevent thousands of visual sightings and UFO videos). And finally, you might argue that the aliens are simply avoiding Colorado. Implausible, but not impossible.

I hope I've shown that my mind is not closed, but merely skeptical. Until someone shows me better evidence, I'm strongly inclined to say that the aliens are likely to be out there, but not here. It's also rather silly that those who believe in flying saucers berate me for not participating in UFO research myself, an accusation they frequently make. Apparently, if I don't investigate the sightings, I cannot make a valid conclusion about alien craft. But that's nutty. I can read a research paper on the discovery of a massive black hole in the center of our galaxy and be able to reasonably evaluate that discovery even though I didn't personally man the telescope. And, by the way, I never insist that these UFO proponents do SETI research, even though they never hesitate to vent their opinions on same.

Finally, and I am remarking more on an oddity rather than a lack of specific evidence, I question why, despite more than a half century of presumed visitation, the sum total of UFO visits have had no significant consequences. This contrasts starkly with visitation experiences on Earth (e.g., the Native Americans). Alien craft may be highballing through the skies, but the Federal Aviation Administration—which is alarmed when ducks fly near airports—doesn't seem worried.

Thousands of people may be hauled out of their bedrooms every night without permission, but the newspapers don't seem to make much of it. Indeed, the only touted impact from all this alien activity seems to be the goofy claim that the government has reverse engineered some of their technology to our benefit. In other words, the extraterrestrials have visited Earth, and all we got was fiber optics.

At least the UFO sightings churn out economic activity that's modest but steady. For example, if you're a member of the Roswell Chamber of Commerce you're probably glad for the navigation error that supposedly brought aliens to the city's outskirts. For T-shirt manufacturers and toymakers, extraterrestrials are as golden as dinosaurs. For television and Hollywood, UFO sightings are endless grist for the mill, providing nonstop tales of incursion, intrusion, and secret government efforts to stack up bodies and saucers at Area 51. In addition, don't discount book and video sales, as well as speaking fees for those who herald this supposed planetary invasion. It's an industry.

All that's just fine. But despite the claims and the commotion, the allegation that aliens visit our planet is, to my mind, unproved. Meanwhile, a different type of search is ramping up: a deliberate, technically sophisticated hunt for signals from extraterrestrials who are still sitting at home.

REVERSE ENGINEERING UFOS

A challenge that Carl Sagan often threw at the feet of those who thought aliens were gadding the globe was to ask the visitors to tell us one new fact about the universe: a fact we could verify with a telescope, microscope, or some other instrument. If the extraterrestrials would only whisper in our ears a tidbit akin to "Hey, there's a hot Jupiter around the star you call HD 434388, with an orbit period of 8.3 days," that would be a prediction we could check. If true, it might compel us to believe aliens really *were* here.

It hasn't happened. But a claim has been made that the aliens have widely advanced our civilization. In 1997, Army Lt. Col. Philip J. Corso co-wrote a book, *The Day After Roswell,* in which he maintained that such perquisites of modern life as lasers, fiber optics, and even transistors were all reverse-engineered from the debris recovered in the celebrated 1947 Roswell UFO crash.

According to Corso, details of these alien technologies were mothballed in warehouse crates until he rescued them in the early 1960s. They were then given to select, trusted people in private industry to develop further. In other words, we can thank the extraterrestrials and Colonel Corso (in that order) for the Internet and iPods.

Investigators have examined Corso's book in detail and found it to be riddled with demonstrable errors of fact involving dates, people, and institutions. But really, that's all pretty trivial when you consider that his premise is less credible than the Easter Bunny.

Corso's claim flies headlong in the face of what we know about the march of progress. Consider, merely as illustration, the history of fiber optics, a technology that the colonel maintains was made possible only by the recovered trash from Roswell. The idea that light could be "piped" from one place to another was demonstrated at least as early as 1841 by Daniel Colladon, a Swiss physicist. The demonstration consisted of setting a large tank of water before his audience. The tank had a spout on the side through which water would surge in a smooth, curved stream, arcing onto the table.

Colladon used mirrors to direct sunlight into the interior of the tank in such a way that it hit the spout at a glancing angle. The light beam shot out the spout, but rather than proceeding in a straight line to the far side of the room, it remained trapped within the water stream, following its curved trajectory. A bright spot could be seen where the water splashed onto the tabletop. This trapping of the sunlight was due to a phenomenon known as total internal reflection. The demonstration showed the basic principle of fiber optics, and Colladon did it more than a century before Roswell.

A contemporary of Colladon's, a French scientist named Jacques Babinet, noted that the same effect could be achieved with a curved glass shaft rather than a stream of water. Babinet used this early version of a fiber optic to facilitate the examination of people's mouths by directing light anywhere he aimed a piece of bent glass.

In the late 1920s and early 1930s, researchers in England, the U.S., and Germany all experimented with transmitting images using bundles of thin glass fibers. An important step in making the technology practical was adding a second coating to the glass fibers (a so-called cladding) that bounced stray light back into the fiber and greatly increased the transmission efficiency. Dutch scientist Abraham van Heel first achieved this feat in the 1950s.

In the 1960s, workers in France, England, Japan, and the U.S. all toiled to reduce the light loss in fibers, primarily by improving the quality of the glass. Eventually this work reached a point where these devices were capable of sending bits of information over long distances. Today fiber optics have overtaken copper wire in long-haul telecommunications.

So much for the brief industrial history. But note the utter lack of discontinuity in the development of this technology. Neither the idea, nor any of the fundamental principles, suddenly emerged in the 1960s, when Colonel Corso said he clandestinely farmed out alien blueprints to selected companies. It's also worth noting that the development of fiber optics was an international enterprise.

In other words, years before Colonel Corso supposedly opened a nondescript crate to discover the secret of optical fibers, all the major ideas had been thought of and implemented by a worldwide assortment of humans. Similar tales could be told for the other technologies that were supposedly made possible by the good colonel's efforts. Alien technology clearly wasn't crucial to producing any of them.

In addition (if you need any additional argument), the whole concept of reverse engineering alien technology doesn't make sense. If extraterrestrials can really fly to our planet, if they have rockets or other transport modes that will bring them from the stars, their level of technology is far beyond ours. How far is "far"? Of course we don't know, but let's be conservative and say "a few hundred years."

OK, do you think we could reverse engineer machinery from a society several centuries in advance of us? That's like giving your laptop to Ben Franklin. He was a smart guy, and he knew more science than most. But despite his best efforts, he would not have been able to reverse engineer your laptop, thereby making spreadsheets and word processors everyday conveniences for his Philadelphia friends.

Next time you marvel at your latest high-tech gadget, you can thank clever scientists and engineers. They were, and are, Earthlings.

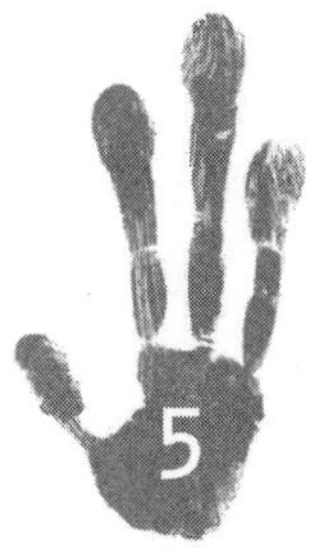

TURNING EARS TO THE SKIES

Tom Pierson, the SETI Institute's chief executive officer, scribbled furiously in one of his signature spiral-bound notebooks. We sat alone in the institute's boardroom at one end of a long, walnut-veneered conference table. C-SPAN was blaring from a television pushed up against the wall opposite, and Pierson was keeping tally of the yeas and nays of Senate members as they voted on a small-time amendment to a major federal spending bill. The amendment had been introduced earlier in the day by Democrat Richard Bryan, a freshman senator from Nevada. He was as focused as a hawk on a vole, intending to kill NASA's SETI program by stopping all project funding for the current fiscal year, about $12 million.

It was 9:30 a.m. on September 22, 1993—early for us, but lunch hour in Washington—and the Senate chamber was sparsely populated. This was good timing for Bryan, and surely gratifying: NASA's tiny SETI effort had been in his sights for two years. The senator had been looking for an issue that would address a persistent federal deficit, one that would play well with constituents and help nail his reelection. In Bryan's view, one could hardly order up a better project to skewer than NASA's search for intelligent alien life. After all, SETI was a small effort (less than 0.1 percent of the total NASA budget) employing only a few dozen engineers and scientists, none of whom lived in Nevada. The social pain of dismantling the project would be small.

Additionally, if SETI were cut, no one could credibly argue that Bryan had killed an effort important to American economic or strategic interests. In addition, and this was a clincher, the whole idea of looking for aliens was easy to caricature as a silly exercise. Few senators knew much about it, and conflating SETI with UFOs and wide-eyed Martians was a tactic that Bryan could use with little risk that his fellow congressmen would call foul. SETI, of course, wasn't trying to find aliens either on the red planet or in the deserts of New Mexico. It was hunting its extraterrestrial prey in situ—light-years away at home. But only the specialists knew that.

When the roll call began, Pierson wasn't worried. He, Jill Tarter, and others involved with the NASA SETI program had made regular visits to Washington to bolster congressional support. In particular, they had repeatedly briefed, and received the backing of, Maryland senator Barbara Mikulski, chairwoman of the Senate's Appropriations Subcommittee on Veterans Affairs, HUD, and Independent Agencies, as well as Utah's Senator Jake Garn, although Garn had left the Senate earlier in the year. I was reassured by Pierson's optimism. After all, in 1993 I was still the new kid in the office. Pierson, by contrast, had been running the institute since its founding in 1984 and was well versed in the Byzantine complexities of NASA funding. I watched the television as the voting continued. There seemed to be a large number of yeas.

Without looking up from his notebook, Pierson said, "We're going to lose." I was stunned as much by his flat tone as by what he was saying. A few minutes later, the vote was complete. It wasn't close: By a vote of 77 to 23, the Senate chose to let Bryan's amendment stand. NASA SETI had taken a gut shot.

I was slightly numb. Pierson left the room and began notifying other members of the institute. Then he began forging a plan to keep the project afloat.

The end of NASA SETI was a big story in the newspapers. "This number has been disconnected," led one *New York Times* report. The cancellation was common knowledge within 24 hours, which prompted an interesting exchange with my next-door neighbor as I pulled into the driveway that evening.

"So, Seth," he ventured, "I guess you're out of a job."

I looked at him with a wan smile: "Well, at least you'll save a nickel on next year's taxes."

The day of the Senate vote, Bryan's office put out a victory press release. The headline was "Great Martian Chase to End?" Its trivializing tone hinted that the senator's staff was unaware that, for a large segment of the public, SETI cast a shadow far longer than its budget and staffing might suggest. After crowing that "millions have been spent and we have yet to bag a single little green fellow," the press release cited other supposed failures of the project, including the sarcastic observation that "not a single flying saucer has applied for FAA approval." Trivially true, and totally irrelevant.

Richard Bryan left the Senate in 2001. His Wikipedia entry suggests that the single most memorable action in his dozen-year congressional career was his campaign against NASA SETI.

In the years before Senator Bryan arranged to vote SETI off the island, the search for extraterrestrial signals had moved considerably beyond Frank Drake's pioneering scrutiny of two nearby star systems. In the 1960s, astronomical studies of natural signals coming from space went mainstream. Radio static from our galaxy, once dismissed as a curiosity without consequence, had garnered the attention of a new generation of scientists and prompted the construction of large antennas sporting sensitive receivers. These included several instruments at the National Radio Astronomy Observatory in Green Bank, West Virginia; the large telescopes at Britain's Jodrell Bank Observatory; antennas of various shapes and sizes built by the Dutch, French, Italians, Swedes, and Russians; and a small horde of telescopes at universities around the world.

Several other antennas that would play an important role in SETI were also erected at this time. As in Britain, wartime radar engineers in Australia were keen to shift from detecting bombers to detecting galaxies, but their attempts to build a significant radio telescope down under were frustrated by a lack of money. Eventually, America's Carnegie and Rockefeller Foundations primed the pump, and in 1961 the 210-foot Parkes radio telescope was completed. Nearly a half century old, it's still the largest single-dish antenna in the Southern Hemisphere. The telescope's large collecting area makes it particularly useful for finding pulsars and mapping the hydrogen in our galaxy. Its finest hour took place on July 20, 1969, when it received the first television images of men on the moon. For this it gained a starring role in the 2000 Australian movie *The Dish.* When I first saw the Parkes telescope standing mournfully alone in Australia's fly-blown sheep country, it looked to me like a prop in a cheesy sci-fi movie.

Puerto Rico's Arecibo telescope, the world's largest single-dish antenna

In the Northern Hemisphere, the largest single-dish antenna is the Arecibo telescope, located 9.3 miles (15 kilometers) south of its namesake town in Puerto Rico. Completed in 1963, it's rather larger than Parkes, with a dish measuring 1,000 feet across, which I reckon could hold four billion scoops of ice cream. It was built in a natural bowl in the Puerto Rican limestone karst, which minimized the amount of earth that had to be bull-dozed out of the way. The island's southern latitude also enabled this instrument to easily target the planets, which cross the skies near the celestial equator.

Operated by Cornell University, the Arecibo antenna has proved extraordinarily productive in studying pulsars, galaxies, planets, and asteroids. For this latter work, it operates as a full-fledged radar, transmitting a one-million-watt microwave signal in the direction of its target. The very large size of the Arecibo dish (equivalent to 373 tennis courts) makes it extraordinarily sensitive—an attractive attribute for SETI experiments. The telescope can detect a signal that impinges on its dish with one-trillionth the energy of an ant taking a single step.

The antenna explosion of the 1960s ushered in a golden age of discovery, comparable to what happened when Galileo first tilted the barrel of his telescope skyward. Galileo immediately picked off some low-hanging fruit: the four largest satellites of Jupiter, mountains on the moon, the crescent phases of Venus, and the vast clouds of stars that compose the Milky Way. As the first on the celestial scene with decent optics, he immediately found unexpected—and in some cases, profoundly important—phenomena.

With their freshly built antennas, radio astronomers four decades ago were also in a position to pluck some low-hanging produce. And they quickly did.

The radio sky was turning out to look quite different from the sky charted by our optical (visible light) telescopes. Sure, the Milky Way could be seen in both spectral regimes, and yes, the Sun churned out a lot of radio static too. But the Sun is just down the road from us, a mere eight light-minutes away. Transport it only a few light-*years* away, and it would subside into silence. In general, the radio sky, which had plenty of "bright" spots, didn't line up with the optical photos. Something other than stars was belching out the bulk of the radio emissions observed with the new radio telescopes.

What was that something? What was making so much radio hiss? To answer this question, the radio astronomers tried to localize some of the stronger radio sources on the sky—that is, they tried to pinpoint the source positions with high accuracy so that their mirror-and-lens colleagues could snap a comparison photograph of that exact part of the heavens. If the optical astronomers found something interesting in their photo—something other than just another scattering of run-of-the-mill stars—they might be able to conclude that the radio emission is coming from . . . whatever. It was like identifying birdcalls in a forest filled with birds.

Obviously, the more accurately the radio astronomers could determine a radio source's position, the more successful this identification effort would be. But radio astronomers were vexed by the same problem that had confronted the wartime radar crowd: They required needle-sharp pointing. So they tried the old radar trick of using microwaves. That helped. But another tactic could sharpen their vision no matter *where* on the dial they tuned: Make the antennas bigger. For wartime radar that wasn't feasible; after all, the antenna had to fit on a plane or a ship. But for astronomy, the situation seemed different. If bigger was better, all they had to do was build something bigger.

It turns out that big has its limits. For a radio telescope operating at the hydrogen emission wavelength of 21 centimeters to be able to "see" as much detail as you can make out with a cheap pair of binoculars, it needs to be 2.5 miles (4 kilometers) across. Such a massive antenna would overhang both the Hudson and East Rivers if set down in midtown Manhattan. That would be an expensive project, not to mention an impediment to traffic.

Fortunately, radio astronomers discovered a way around this dilemma—rather than build one large antenna, you can build lots of little ones. If you then combine the signals collected by this multi-element array, you can mimic the resolution—the ability to see detail—of an antenna as large as the distance between the farthest two members of the array.

This technique—first put into practice at Cambridge University's radio observatory—has been used over and over by radio astronomers. They've built ever larger arrays—at first a few hundred feet in size, then tens of miles. Today, it's possible to electronically hook together antennas that are a continent or even an ocean apart. Typical resolution for these Very Long Baseline arrays is a hundred times better than any mirror-and-lens telescope, equivalent to mapping a sand grain in San Francisco from New York City.

Armed with such arrays, radio astronomers had the ability to make the measurements that would finally lead to the identification of the mysterious, bright radio sources they had discovered years earlier. Some sources turned out to be relatively nearby objects: dead stars that had either exploded (supernova remnants) or collapsed (pulsars). Others were far more distant—galaxies that had immensely energetic activity in their cores (quasars), tumult that was eventually traced to mammoth central black holes. In addition, the ability to extract fine radio detail led to the detailed mapping of our own galaxy, as well as other nearby galaxies. Radio astronomy was on a roll.

My own research career began with a simple radio array (it had only two antennas) situated at Caltech's Owens Valley Radio Observatory.* For weeks at a time, I would reside at this desert outpost in a narrow valley east of California's Sierra Nevada. For my thesis I studied galaxies at the hydrogen wavelength in order to puzzle out their motions. Every day I recorded the faint radio hiss emanating from such

**In an ironic coincidence, and a matter of personal pride, the telescopes I was using at the Owens Valley, as well as others throughout the country, including the first antenna at the Hat Creek Radio Observatory (now the home of the Allen Telescope Array), were the recipients of major financial support from the Electronics Branch of the U.S. Office of Naval Research. My father, Arnold Shostak, was the head of this branch for several decades.*

cosmic all-stars as the spiral galaxy NGC 2403 and its smaller cousin, NGC 4236. To my surprise, these mammoth stellar conglomerations seemed to rotate too quickly in their outer regions. This would turn out to be an important clue to the existence of what's now called dark matter.

Radio observatories of that era didn't have large staffs, and students were expected to do everything themselves. That included operating the telescope through the night. The work was lonely and routine: Every few minutes I flipped a few toggle switches to change the positions of the antennas, and checked that the electronics were still working. Then I went back to whatever reading matter was distracting me for the evening. One night I was deep into a new book by Iosif Shklovskii, a Soviet physicist, and Carl Sagan: *Intelligent Life in the Universe.* Their small treatise was the first comprehensive study of how modern technology could allow us to find creatures on other worlds.

Paging through the book to the background whir of muffin fans and the distant growl of the tracking motors, I became aware that the simple antennas beyond the control room windows had the power to do more than study the natural radio cacophony of space. They could—at least in principle—also be used to find other inhabitants of the cosmos.

Imagine: Small aluminum-and-steel antennas and a bit of electronics—hardware that could fit on a mall parking lot—might sense the presence of strange beings hundreds of trillions of miles from Earth.

By the 1970s, I had taken a postdoctoral position at the National Radio Astronomy Observatory. There I used single-dish antennas to pursue my studies of galaxies. As a consequence, I was often at the 300-foot antenna in Green Bank. During the occasional breaks in the schedule—10 or 20 minutes during which no suitable galaxies were accessible to the giant dish—I would often fill the slack time by pointing at nearby star systems in an improvised SETI experiment. This was like feeding pocket change to a slot machine on the off chance of hitting the jackpot—in this case, an artificial signal that would come booming in and make my day.*

I didn't find any signals, and I didn't publish. But I wasn't the only one. In the decade that followed Project Ozma, many radio astronomers devoted occasional free telescope time to their own "Ozmas." These informal searches were also shots in the dark, but a few astronomers went beyond viewing SETI as cheap caulk to fill in the gaps between more serious observing projects. These researchers went to the radio telescopes for *deliberate* SETI observations, carefully choosing targets and writing up their (negative) results in a way that would put numerical limits on certain types of signals. But their efforts were akin to an engine firing on only one cylinder: The equipment they used was built for radio astronomy, not for SETI. Consequently, these pioneering hunts for alien transmissions were necessarily slow and insensitive.

Nonetheless, when viewing this early work in retrospect, one could at least credit it with establishing an important fact: The sky was not—repeat, *not*—filled with strong, persistent radio signals from other worlds. This was unlike the circumstance in Europe at the end of the 15th century, when any ship sailing west for a few months was guaranteed—*guaranteed*—to hit a

---

**The National Radio Astronomy Observatory's 300-foot telescope, as it was aptly if unimaginatively called, collapsed due to a faulty gusset plate in November 1988. The tabloid* Weekly World News *reported that the large antenna's demise was the deliberate work of space aliens who were fed up with being the object of its scrutiny.*

previously unknown continent and rack up a major find. Sure, bobbing slowly across the North Atlantic was dangerous work, but it was straightforward. Discovering ET wouldn't be.

Fortunately, when the going got tough, the tough got organized. Since SETI wasn't likely to uncover subtle extraterrestrial signals by operating in "garage mode"—quick peeks by astronomers who were casually filling gaps in their observing schedules—its adherents desired a more systematic project. The question was, who would mount this effort? Obvious candidates were radio astronomers. After all, they had been dabbling in SETI ever since Drake's 1960 experiment.* But the small sizes (and budgets) of most university astronomy departments augured against anyone making a major enterprise out of this niche research area.

In the end, a medical doctor started the first truly serious program to hunt for alien radio signals. John Billingham, an English physiologist who had worked on protective suits and other space flight technology for both the Royal Air Force and NASA, launched SETI as a purposeful enterprise. In September 1965, Billingham had been lured to the Life Sciences Division at NASA's Ames Research Center, located in the then agrarian area south of San Francisco that would soon ripen into the Silicon Valley.

"It wasn't long before I was chief of the Biotechnology Division there," recalls Billingham today. "And over the years I got very intrigued with the people on the third floor of the Life Sciences building. They were part of something called the Exobiology Division. We used to meet and chat over lunch."

***Of course, astronomers weren't the only ones who might have an interest in a bigger SETI project. Such a program might have appealed to a daring group of biologists.**

Billingham was being exposed to the nascent ideas of SETI.

"It dawned on me that there was potentially a big gap in NASA's space program. The Exobiology Division was worrying about the chemistry of life, and even looking for life within our solar system. But what about *intelligent* life around *other stars?*"

Shortly thereafter, Billingham and Jim Adams, a professor of mechanical engineering at Stanford University, cooked up a plan to examine the SETI topic in a joint summer engineering design study. They intended to herd two dozen academics into NASA offices and pay them to spend three months working out what sort of telescope was needed to make an effective search for intelligent life. (It was unclear whether "effective" included outcomes other than "successful.") The study participants were also charged with estimating the costs of any necessary equipment, including new radio telescopes.

The summer design study was a natural fit to NASA's brief. After all, the space agency had been seduced by the possibility of life in space for years, particularly on Mars, and it certainly had experience in building complex hardware. It also had budgets commensurate with big projects.

Billingham recruited a co-director for the study: Barney Oliver, an electrical engineer who was the director of research for the well-known technology company Hewlett-Packard. Oliver, an imposing man both physically and intellectually, had visited Frank Drake in 1960 at Green Bank and had written about SETI. In any gathering of technical types, Oliver was the 600-pound gorilla. The other members of the study group, all 22 of them, hailed from engineering faculties at American universities.

"I commandeered a spare floor in the space sciences building at Ames," says Billingham. "We constructed little offices for all the professors, and had a gorgeous, glorious summer."

Billingham's brain trust drilled down into the details: What sort of signals might be most easily transmitted from other star systems and detected on Earth? What equipment would be necessary to find these signals, and how could that hardware be assembled? Eventually they devised plans for an imposing new instrument: an array of antennas, each a hefty 328 feet in diameter, that would be aimed at nearby stars, and whose receivers would comb the microwave radio spectrum for signals. The number of antennas would expand depending on progress—a few antennas would be built at first, and the tally could be increased to an ambitious several thousand. The price tag, which depended on exactly how many of these electronic ears were deployed, ranged from one to thirty billion dollars. After the summer sessions ended, Oliver spent more than a year putting together a comprehensive write-up of the study. He named it after the one-eyed giant of Greek mythology, Project Cyclops.

It's hard to overestimate the influence wielded by this 1970s publication. Even today its wake rocks the boat of anyone who hopes to find company among the stars. The Project Cyclops report laid out a bold plan, a daring plan, and a plan that has gone unfulfilled. Written during the euphoria of the Apollo moon landings, it anticipated that the enthusiasm for space exploration would continue to swell, including pursuit of one of its most lustrous trophies: the discovery of intelligent life.

Popular interest seemed to mirror that of the Cyclops participants. In 1978, Carl Sagan authored a classic paean to the SETI enterprise for *Smithsonian* magazine, detailing both the experiment's history and its hopes. Sagan wrote,

"It is difficult to think of another enterprise which holds as much promise for the future of humanity." He also said that detecting another, more advanced cosmic society would prove that it's possible to survive "technological adolescence"—when any society develops massively destructive weapons. If aliens could manage to survive the invention of nuclear bombs, we might take heart about our own future. At this time Sagan was beginning work on his mammothly successful TV series, *Cosmos,* which first aired in 1980. Space, and its putative inhabitants, were as fashionable as miniskirts.

Billingham's team arranged for artists' renditions of the Cyclops antennas, a proud but possibly counterproductive gesture. The picture of a full-up Cyclops antenna farm—a thousand telescopes crowded shoulder to shoulder like poppies in a field—looked expensive, and it was. But this maxed-out array wasn't really the thrust of the Cyclops recommendation. Rather, and as the report carefully stated, the array was to be built incrementally. After all, no one could predict what degree of sensitivity was required to find an alien signal. So Cyclops would begin small and expand in size only as necessary. If a hundred antennas weren't adequate to pick up ET's transmission, more would be added. Expenditures would always be no more than the minimum required to make a detection, assuming that a detection was ever made.

Cyclops proved to be similar to Lyndon Johnson's Great Society—a beautiful blueprint, but an unconstructed edifice. Its antennas are, to this day, still no more than sketches in a book. But the ideas developed during the summer of 1971 continue to shape SETI research.

As an example, the study team recognized that the 21-centimeter hydrogen wavelength that was chosen as a cosmic hailing channel by the first SETI scientists was only a twist

Artist's conception of full-up Cyclops array

of the tuning knob away from the 18-centimeter emissions of another natural noisemaker, the hydroxyl radical, or OH: a simple two-atom marriage of hydrogen and oxygen. This proximity was both fortuitous and suggestive. The Cyclops scientists recognized that the small spectral stretch between 21 and 18 centimeters would be known to any advanced society—be it terrestrial or alien. This special part of the radio dial was arguably a dead-obvious tuning choice for those wishing to ping other galactic inhabitants. In addition, hydrogen and hydroxyl are the components of water, the one essential ingredient for life (at least, life as we know it). As Oliver wrote, this small band of frequencies "was an uncannily poetic place for water-based life to seek its kind. Where shall we meet? At the water hole, of course!"

Today, more than three dozen years after Cyclops, the frequency band between 21 and 18 centimeters remains the favorite spectral hunting ground for SETI practitioners. Some may listen at other spots on the dial, but *all* listen in the water hole.

JOINING THE TEAM

By 1974, Billingham had snagged his first serious funding for continued NASA involvement in the search for sentience beyond Earth. But the money was still small, and the space agency fielded workshops and meetings, not antennas. At one such gathering in Puerto Rico, the participants decided to change the name of the program. Until 1976 it had been called CETI, or Communication with Extraterrestrial Intelligence. The new sobriquet replaced Communication with Search, and the attendant acronym morphed from CETI to SETI. Everyone understood the motivation. The aliens were undoubtedly many light-years away, and *communication* would be tedious. Searching, on the other hand, might turn up a signal before tomorrow's breakfast.

The NASA SETI effort was gaining traction. By 1983, Barney Oliver had retired from Hewlett-Packard, and Billingham hired him to be deputy chief of the SETI Program Office. In that same year, Tom Pierson, an administrator at San Francisco State University, made a persuasive pitch to Billingham and his new hire, Jill Tarter, about the benefits, both organizational and financial, of creating a nonprofit entity to manage the NASA SETI effort. By the fall of 1984, the requisite paperwork was completed, and the SETI Institute became reality. The founding of this organization was seen at the time as merely an organizational maneuver that would reduce loss of NASA research monies through high overhead rates. But a decade later, when the NASA SETI program was canceled, the existence of this nonprofit would save America's largest SETI program. Government funding for SETI might drain away, but the organization would survive. Eventually, Pierson's brainchild blossomed into a wide-ranging research organization that encompasses much more than just its namesake SETI searches. Today, institute scientists hunt for microbes on Mars and the moons of the outer planets, and investigate how life began on Earth, among other things.

I didn't witness any of these developments: I was working an ocean away in the Netherlands, at the state university in Groningen, a small town that no one's ever heard of, and where tourists never go. I had moved there in 1975, to do astronomy with the then new Westerbork Synthesis Radio Telescope. This boasted an array of 12 antennas, spaced east-west along a half mile of Dutch real estate about 30 miles (50 kilometers) south of Groningen. At the time it was the most advanced radio array in the world.

You might wonder why a small country like Holland had forged ahead of any of its European cohorts in building such a sophisticated research device. But the Dutch have a long history of astronomical inquiry. They became interested in radio astronomy early, and as soon as the Second World War ended, they rebuilt some abandoned radar antennas deployed by the Nazis near the North Sea, the so-called Würzburg dishes. These were erected to warn the Germans of Allied air attacks, but Dutch radio astronomers refitted the dishes and used them to make the earliest charts of the Milky Way's structure. Swords into plowshares.

The postwar work eventually led the Dutch, under the leadership of one of the most accomplished astronomers of the modern era, Jan Oort, to construct Westerbork. By the 1970s, the University of Groningen was looking for researchers to use their new telescope. They began importing radio astronomers from around the world. This campaign was so successful that by the time I arrived, only a minority of the faculty was fluent in Dutch. I soon began using the Westerbork array to map more galaxies.

But I also was developing a growing fascination with SETI. I would occasionally give a talk or two, or write an article for popular magazines, on the subject. My interest was long-standing, dating back even further than the epiphany I'd had while observing at Owens Valley. As a middle school student, I had read a few books about life in space—mostly about flying saucers. It was a subject that, like dinosaurs, infected lots of kids.

This modest, mostly uninformed interest eventually prompted me to dip my toe in the SETI waters. In 1981, Jill Tarter took a short sabbatical in Groningen. This was a small step for a female astronomer, but—as it turned out—a big step for my future. I hadn't conducted any SETI experiments other than the occasional quick looks at nearby star systems during my observing programs at Green Bank. But Tarter's presence rekindled my interest, and I soon made her acquaintance. Within weeks we had penned a proposal to use Westerbork to search for artificial signals coming from the center of the Milky Way.

Why the center? That's because the galactic nub is unique. Not just special, but *unique.* It's the only place in the Milky Way that everyone will know, and everyone will study. If our galaxy houses truly advanced societies, perhaps one among them has installed a radio beacon at this singular spot, confident that other sentient beings will routinely look there.

We were given four hours of telescope time: a pittance. But then again the Dutch astronomical community generally regarded SETI as wishful thinking.

"We don't expect you'll find any signals from ET," the dean of the department explained, "but maybe you'll learn something interesting about the behavior of the telescope."

Our little experiment, called SIGNAL (for Search for Intelligence in the Galactic Nucleus with the Array of the Lowlands, a slightly stumbling acronym conjured by Tarter), failed to detect a galactic beacon in the Milky Way's downtown districts. It also failed to reveal any outstanding problems with the telescope, which undoubtedly disappointed the dean. But it hooked me on the excitement and potential importance of this effort.

By 1990, I was once more living in the States, this time in the Silicon Valley, where I had moved to join my brother Rob in a start-up software company. I left Holland reluctantly: I was happy living and working there. But my family and my new wife, Karen, pleaded for a return to the country of my birth and my relatives. Rob made that return possible by offering me a job with his start-up.

Unfortunately, the company went ventral side up after 18 months, and I was left scrambling to earn grocery money. In a life-changing fortuity, I attended a dinner party in Berkeley. There I ran into Jill Tarter, and shortly thereafter Billingham and Pierson were calling to see if I was interested in a job at the SETI Institute. I was.

Given the unusual nature of this work, I'm frequently asked if such an uncertain endeavor can make for a satisfying career. The answer can be as lengthy as a polite regard for the inquisitor's time permits. But—and I realized this from the beginning—it's a privilege to work on a problem of such consequence. The question of whether intelligence is a phenomenon specific to Earth is one that many of our predecessors must have asked. After 200,000 years of human existence, we've reached a point in our understanding and skills that might realistically lead us to an answer. There are no guarantees, but life is hardly awash in guarantees. I could have been an accountant—a vocational choice that would have been more lucrative and secure. But does an accountant have the possibility of changing humanity's collective mind forever?

THE LISTENING BEGINS

In the first years of my presence at the SETI Institute, the team—which had grown to several dozen scientists and engineers—was busy gearing up for a major, systematic radio reconnaissance of the skies. Billingham's group at NASA Ames now had an opposite number 400 miles (644 kilometers) away at the Jet Propulsion Lab in Pasadena. JPL had become involved with SETI in 1977, mostly at the insistence of the lab's director, Bruce Murray. Both groups were designing and building sophisticated analyzers capable of simultaneously monitoring tens of millions of radio channels. After all, even though the water hole seemed like a good bet for listening, a few hundred million frequency slots lay within that celebrated spectral window—any one of which might carry a signal. To speed up the search, we needed to simultaneously listen to as many channels as possible.

The two groups were also refining their strategies for sniffing out ET's presence, and each adopted a different approach. The Ames team wanted to elaborate on Drake's pioneering experiment: Simply point the antennas at nearby star systems, preferably those anchored by stars like the Sun. These systems, one presumed, would be the most favorable neighborhoods for the emergence of intelligent life. This conservative strategy was inarguably reasonable: Look for aliens in solar systems.

The JPL crowd thought such a targeted search was too restrictive. They preferred to avoid making *any* assumptions about where the aliens were living. How could we divine the sorts of habitats that might appeal to beings millions of years beyond us? So JPL opted to sweep the entire sky (or at least that part of it accessible to their antenna in the Mojave Desert). The trade-off for examining this greater swath of celestial real estate was reduced sensitivity in any particular direction. Extraterrestrial signals would have to be more powerful to ring bells on their receiver. The JPL scheme was called, unsurprisingly, a "sky survey."

The NASA SETI experiment—both the targeted search and the sky survey—would have to proceed without the covey of new antennas envisioned in the Project Cyclops report. The political situation had changed. Once the last Apollo astronaut departed the moon—leaving both footprints and the Soviet Union's lunar aspirations in the dust—the public felt satiated with space exploits. Building a vast, multibillion-dollar radio telescope to hunt for extraterrestrials was a nonstarter. Ergo, the researchers would scale back. Their revised game plan was to do what SETI scientists had always done: Search with existing antennas designed and built for radio astronomy.

So how did we do that? I'll describe, as an example, the targeted search. The Arecibo Observatory allotted several weeks per year for SETI. When it was our turn on the telescope, we'd trek down to Puerto Rico (an all-day trip from California) and, after checking into the commodious observatory dorms, set up our equipment. This last consisted of several parts. The most complex was the digital receiving electronics—racks full of custom circuit boards that would, when fed the cosmic static collected by the antenna, sort it into 28 million individual channels, each a very narrow 1 Hz wide. (For non-radio engineers, that's about five million times narrower than a standard television channel.) Once the signal had been spectrally sliced and diced, it was ingested by the signal analysis system, a combination of hardware and software that combed through those millions of channels for a signal, a "hit." Our signal analysis system was the digital equivalent of Jodie Foster's ears, although it was both more sensitive and more capable (28 million times more capable, in fact). And unlike Jodie, our equipment could also detect transmissions drifting up or down the dial, or slowly pulsing on and off.

If that's not enough techno-pleasure for you, allow me to mention the verification software. It would ferret out any signals and subject them to automated tests to decide if they were clearly terrestrial in origin—for example, if they were at a frequency at which we had already found man-made interference. This process constantly revealed phonies: At Arecibo, signals were detected nearly every ten seconds. Needless to relate, terrestrial interference was quickly yanked off the processing conveyor belt.

Finally, we had additional software to select which star system to examine and to drive the telescope.

It all added up to a pretty complex helping of computers, custom circuit boards, and analogue electronics. To keep from having to take it apart and reassemble it every time we moved to another radio telescope, we packaged most of the hardware into a lovely white shipping container, outfitted with heavy-duty air-conditioning. This made the hardware modular, and easy to transport from place to place. In addition, the big steel box shielded all the electrically noisy digital electronics within, lessening the chance that, at the telescope, we'd pick up our own equipment's emissions and mistake it for ET. That would be embarrassing.

Weeks before we started observations, the trailer-size "mobile research facility" (or MRF, pronounced murf) was already parked outside the control room building at the observatory. It impressed casual visitors with its bulk, its loud air-conditioning noise, and its colorful NASA logo. It also promised to usher in an era of serious searching for extraterrestrials.

On October 12, 1992, exactly 500 years after Columbus first splashed onto the beaches of the New World, NASA initiated its two-pronged search for signals from other worlds. Near Mojave, California, the space agency's Deep Space Network antenna began ranging across the sky, and in Puerto Rico the Arecibo radio telescope started its one-by-one scan of a thousand nearby star systems.

The hunt was on. A stunning voyage of exploration cast off the hawsers and pulled away from the docks. Initial progress was halting, as the equipment sputtered and glitched. Data were dribbling in. Then, just as this shakedown cruise cleared the breakwater, the scientists received the dismaying order to come about. Senator Richard Bryan had killed the project.

RISING FROM THE ASHES

As September 22, 1993—Black Wednesday—ended, members of the SETI Institute's upper echelon were scrambling for a life raft. NASA was now obliged to shut down its funding for the search, as well as end our use of the agency's 108-foot Mojave antenna.

There was a possible way out, however, and it derived from the fact that Pierson had enrolled most of Billingham's team in the SETI Institute. The existence of this entity was the key to continuing the search, not as a NASA project, but as the privately funded effort of a nonprofit research organization.

The institute started talking to fund-raising experts. Surely donating to a project as exciting as the scientific search for cosmic intelligence would appeal to people of means. Barney Oliver called folks he knew in the high-tech industry. He began with his former employers, William Hewlett and David Packard. Each made a five-year pledge of one million dollars annually, or ten million dollars total. Within hours, he had elicited the same from Gordon Moore (co-founder of Intel Corporation) and Paul Allen (co-founder of Microsoft). These pledges were drawn on personal, not corporate, bank accounts.

In retrospect, this early, easy success at garnering monies for SETI was misleading. If Oliver could raise $20 million in an afternoon, we foresaw little danger that SETI would ever be threatened by a lack of funds. Time and experience would show how naive this assumption was. The personal checks Oliver corralled with a few phone calls were written because of long-standing relationships. I honestly think that had he asked for donations to his fishing club, his friends would have agreed to that, too. Barney Oliver had earned and enjoyed enormous esteem.

Now that adequate funding was guaranteed for at least five years, plans were drawn to relaunch parts of the prior NASA program. The all-sky survey favored by JPL was dropped in favor of continuing the targeted scrutiny of a thousand nearby stars. In addition, arrangements were made for the long-term loan of the NASA SETI equipment. The big trailer was ours to use.

Within three years, SETI had rebounded as a "private enterprise," and a pared-down search was under way. The grandiloquent speeches of Columbus Day 1992 were reprised in an outdoor dedication ceremony on the grounds of the 210-foot Parkes radio telescope west of Sydney, Australia.

On February 2, 1995, the new targeted search was christened Project Phoenix, a restrained bit of whimsy alluding to its rise from the ashes of the canceled NASA effort. As the crowd sweated in the heat of an Australian summer day, the antenna motors revved, and the Parkes dish twisted in the direction of our nearest stellar neighbor, Alpha Centauri.

Project Phoenix continued for nine years, and was serially conducted on three primary instruments: the Parkes dish, the 140-foot radio telescope in Green Bank (where the false alarm of 1997 was picked up), and the 1,000-foot Arecibo reflector. It was, and at this time remains, the most comprehensive search for radio transmissions from nearby stars ever conducted. It combed an unprecedented stretch of the radio spectrum: 1,200 to 3,000 MHz, a swath more than ten times as wide as the water hole.

Phoenix examined about 750 star systems—stars that were, in the main, similar to the Sun in terms of brightness and size. It failed to find a signal that could be attributed to an alien transmitter.

To much of the populace, Phoenix's lack of success suggested that SETI was a hopeless endeavor. Either the aliens weren't out there, or the schemes used to detect them were faulty, in their premise or their execution.

Let me assure you that this conclusion was unwarranted and naive.

SCRATCHING BENEATH THE SURFACE

The star systems monitored as part of Project Phoenix were typically less than 150 light-years away. While 150 light-years would be an astonishing tally on your car's odometer (roughly 1,500,000,000,000,000 kilometers, or a thousand trillion miles), it's merely a few strides into our stellar backyard. The diameter of the galaxy is close to 100,000 light-years. In other words, if the Milky Way were the size of a dinner plate, Phoenix's reconnaissance range would be a pencil dot.

Reckoned as a fraction of the total number of stars in the galaxy, the sample scrutinized by Phoenix is likewise trifling. Imagine a haystack taller than you and bulky enough to block the view of your car. If that mound of piled-up pasturage represents the galaxy, then Project Phoenix examined less than a teaspoon's worth of hay.

So, contrary to common belief and the occasional misguided declarations of pundits, it is neither surprising nor significant that this first systematic search of the galactic haystack failed to find any needles.

But how should SETI proceed? The prospects look daunting. While no one knows how many broadcasting civilizations infest the Milky Way (see SETI's Famous Formulation: The Drake Equation, page 106), we could, just as an example, consider the implications of Carl Sagan's optimistic estimate. He guessed that a million or more such voluble worlds exist—a million or more places in the galaxy are pumping out emissions we could detect. If so, then one in a few hundred thousand stars will have a planet from which we might expect a signal. When will we find one? That question is the same as asking how many years it will take to examine that many stars. At the rate at which Project Phoenix sifted the sky, we will labor for a *millennium* before discovering ET. That's a long time to be drumming up funding, let alone maintaining one's professional interest.

But it might be worse: Sagan's sunny estimate of the number of transmitters was on the high side, as such guesses go. It's 100 times greater than that ventured by Frank Drake, who feels the number of signaling societies might be only 10,000. If Drake's guess is closer to the truth, then SETI researchers need the patience of Job. They shouldn't expect to bag their first prey before the year 100,000.

That's depressing enough to discourage even Dr. Pangloss. Fortunately, these somber prognostications evaporate once you realize that the SETI enterprise improves with time. The search becomes ever faster, a fact often missed by those who are only casually acquainted with these experiments. For most members of the public, the perception of SETI has been shaped by entertainments such as *Contact* or *The X-Files*. They imagine scientists sitting around dutifully monitoring static from dusk to dawn, an activity as monotonous as *Bolero*.

At one time we really did do that. Back when Frank Drake was busy with Project Ozma, he literally sat at the telescope waiting to see what happened. The signal from his single-channel receiver was traced out on a chart recorder, making a paper record much like a seismograph.

A dozen years later, better electronics for monitoring hundreds of simultaneous channels were used for SETI and, a dozen years after that, the number of channels had increased to thousands. Now the tally is tens of millions or more. We've long since abandoned using chart recorders, earphones, or any other analogue devices as monitoring devices. Computers were long ago given the job.

In fact, today almost every facet of signal processing for SETI is digital. Digital electronics, not analogue filters, slice and dice the incoming cosmic static into those myriad channels, and then sift through them looking for persistent signals. Until the 1990s, the computers used were off-the-shelf models, programmed as required. Since then all SETI projects have used custom-made digital circuitry, which, while more painful (and more expensive) to put together, can be faster than a general purpose computer. "Faster" means that more channels can be examined simultaneously, thus speeding up how quickly a star system can be checked out.

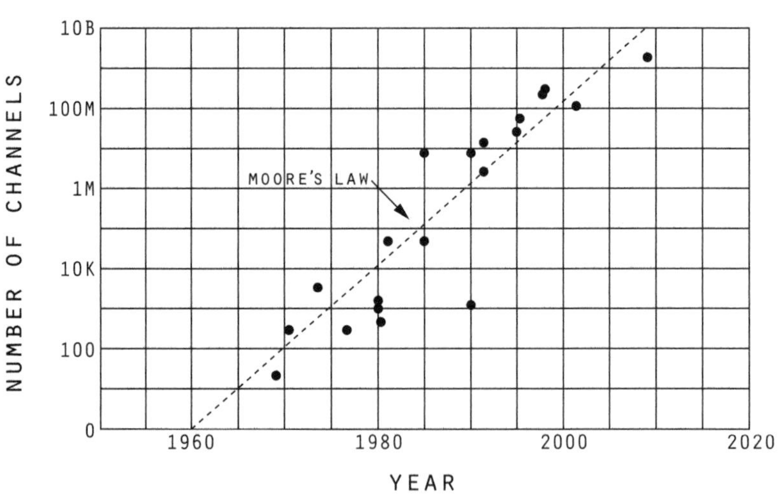

An index of SETI speed, the number of channels simultaneously observed, is following Moore's law (straight line).

Yet as the computer chip revolution snowballs, this reliance on custom hardware is probably ending for SETI. The best off-the-shelf computers are now nearly as fast as what we can build, and they are more reliable. The effort now expended on designing and building specialized circuit boards will be turned to programming. The right computer code can transform a general purpose computer into a SETI listening post.

You can now appreciate one of the more delicious incongruities of the movie *Contact.* In the film, Jodie Foster sits on the hood of her car in the New Mexico desert, with earphones playing celestial static to her bored brain. But in the technically correct version, she would be wearing *tens of millions* of earphones (or using a computer). Surprisingly, when this fact was pointed out, Warner Brothers refused to give it much consideration. Presumably the production company worried that wearing millions of earphones would crowd the shot.

The point is, since digital computing is the man behind the curtain for so many aspects of SETI, faster computers beget a speedier search. And as you've surely noticed, computers are getting faster.

For decades the improvement in computer speed has followed something called Moore's law, after the aforementioned Gordon Moore, a chemist and co-founder of Intel Corporation. The "law" boils down to a simple fact: The amount of compute power you can buy per dollar doubles every 18 months. You might think this exponential improvement in performance is a technical phenomenon, and of course in some sense it is. But Moore's maxim is, at root, an economic law. Most people are willing to buy new computers every three to five years, following the same replacement cycle they usually adopt for automobiles. Ergo, the chip manufacturers improve their product quickly enough that, after this number of years, the new model is a substantial improvement on the one you have. Not that the old one has worn out, or been scratched, dented, or dinged—computers aren't exposed to the elements, potholes, or parking lots. But you still want a new one because it has significantly more capability.

Although many technologies improve rapidly, few have sustained a decades-long *exponential* performance increase. The steam engine didn't, and automobiles don't. But computers are different. They're involved in designing (and building) their successors. Such a circumstance leads to geometric improvement.

So SETI performance is tightly tied to computer performance. In the figure on page 182, I've graphed one measure of the speed of SETI searches, in this case the number of channels simultaneously monitored. A glance at the plot shows that, on average, this number has been rising ever since Project Ozma. A closer scrutiny will show that the rapidity of the search has been climbing *exponentially,* and in good agreement with Moore's law. On the whole, SETI experiments double in speed every 18 months. This is why all the dour estimates of how many millennia will tick by before a signal is found are bilge and balderdash.

THE ALLEN TELESCOPE ARRAY

At the end of the 1990s, the SETI Institute brought together about 40 scientists and engineers in a modern reprise of Billingham's earlier summer design study. Their task was to chart a road map for the projects SETI should undertake in the first two decades of the 21st century. Bowing to precedent, they eventually produced a book that summarized their deliberations: *SETI 2020.* One of the most important conclusions directed SETI practitioners to build a radio telescope designed from the pedestal up to be optimized for a search. This would, the task force realized, shift SETI into high gear.

It may occur to the alert reader that this recommendation smacks of Project Cyclops, itself a proposal that submerged under its own weight. But technology has advanced in the decades since NASA's pioneering study. For starters, digital circuitry now allows for the inexpensive construction of those multimillion-channel receivers Jodie Foster wouldn't deign to monitor. The task force also heard the inexorable rumbling of Moore's law and realized that any new instrument should be designed with the anticipated performance increase of computers in mind. Digital manipulations are cheap, and getting cheaper. But the technology changes aren't all caused by the bit-chip revolution. The sensitive amplifiers mounted at the focus of any radio telescope—the analogue hardware that boosts the level of the incoming waves—have come down in price by about a factor of a hundred since the early 1970s.

This technical tidbit is something that, like as not, has escaped your notice. But its consequences for telescope designers is this: The cheapest way to build a truly powerful instrument is to fashion it out of lots of relatively small antennas. You can appreciate the logic by considering the following thought experiment: You're charged with building a radio telescope with a given collecting area, say an acre of aluminum mesh for catching signals from the sky. But instead of building one large antenna, you opt to build two smaller ones, each with half the collecting area. The acreage of each dish is now reduced by a factor of two, but the amount of steel in each is reduced by a factor of nearly three. So the two dishes together weigh less than the single larger one.

It's an exercise left to the reader to extend this argument to yet smaller dishes, but clearly the advantage of doing so accumulates. The bottom line is the bottom line: It's less expensive to build many small antennas than one big one, even when the total collecting area is the same.

In the past, this approach—which clearly minimizes the amount of fabricated steel and aluminum required—would falter on the fact that the amplifiers at the focus cost about a million dollars per antenna. More telescopes meant more amplifiers, and a simple cost-benefit analysis proved that maximizing bang for buck involved building an array with antennas that were between 75 and 100 feet in diameter.* But with today's cheaper receivers, the optimum antenna size has shrunk to about 20 feet, and the total cost (for a given sensitivity) is a factor of three less than it once was. Same performance, one-third the price tag.

Today, you can find the practical consequence of all this theorizing in the lava-strewn pastureland north of Mount Lassen, California, 311 miles (500 kilometers) from San Francisco. Here, surrounded on all sides by imposing topographical relief, is the Allen Telescope Array, the progeny of the *SETI 2020* task force. The telescope bears the name of its principal benefactor, Paul Allen, who provided most of the $30 million capital cost of designing and building the first 42 elements of

---

**The poster child for this erstwhile economic reality is the Very Large Array in New Mexico, dedicated in 1980, and consisting of 27 antennas, each a hefty 85 feet across. Incidentally, although the VLA has appeared in several movies, most notably* Contact *and* Independence Day, *it doesn't possess the digital electronics necessary to slice the radio spectrum into extremely narrow channels—the type of strategy that most SETI researchers think is optimal for finding an extraterrestrial signal. As a consequence, despite its movie cachet, this famous array has never been used for SETI.*

an array that will eventually contain 350 antennas, scattered in a semirandom pattern across a half mile of real estate.

The Allen Telescope Array can perform many research tasks faster and better than competitive radio telescopes. It's a seductive instrument, even aside from its abilities to search for extraterrestrials. Consequently, its construction and operation have been a joint project of the SETI Institute and the University of California at Berkeley's Radio Astronomy Lab. Observing will also be a joint effort, since the two organizations can time-share. For example, while the Berkeley astronomers map a galaxy, the SETI researchers can examine nearby stars that happen to be in the same field of view. That field of view is extraordinarily wide, typically many degrees across. In other words, the array is a wide-angle radio lens, and there are always nearby stars within its purview no matter which direction it's pointed. In this sense it's like looking for field mice while tagging along on an elephant hunt: No matter where your companions search for pachyderms, there are likely to be mice in the vicinity.

If we consider only the SETI capabilities of this instrument-in-the-making, we find the following benefits:

1. The array's nonstop accessibility ratchets up the speed of the search by about a factor of ten. No longer is SETI an on-again, off-again activity, condemned to explore the unknown waters of the cosmic ocean in rented ships.

2. Rather than focusing on one star system at a time, the array can simultaneously examine several. With sufficient computer power, that number can increase to dozens, hundreds, or possibly even thousands. This provides another very significant speed boost.

3. Instead of being restricted to a small twist of the radio dial near the water hole, the new instrument can sample

frequencies from approximately 500 to 11,500 MHz. This stretch of spectrum is six times greater than that examined by Project Phoenix, and the array's new digital processors can sift through it faster.

All of these rather esoteric technical advantages boil down to a profound step forward, one that could truly affect future history. As we've noted, if the hunt for ET continued at the pace set by Project Phoenix, it would take somewhere between 1,000 and 100,000 years before a signal was found. But the new array changes the odds. If we can take advantage of Moore's law to continuously improve this instrument, we'll stop the frustrating exercise of sampling the cosmic haystack with demitasse teaspoons; we'll have a soup spoon. Within a few years, that will become a ladle. In a decade or two, we'll be flying through the fodder with a shovel.

A straightforward calculation shows that by the year 2030, the Allen Telescope Array could check for signals in the direction of a million or more star systems. That's enough to offer success if Drake is correct. If Sagan's right, a signal will be found sooner. In other words, either we will discover evidence for ET within the lifetime of the present generation or we've erred badly in our presumptions.

You could argue—as do many of my colleagues—that the numbers of broadcasting societies estimated by Sagan and Drake are only guesses. Consequently, this simple prediction—namely that a success will occur within a generation—is wobbly at best and bravado at worst. I hear this in the hallways a lot.

Indeed, it's fashionable among some of my colleagues to claim that SETI should be compared to building Gothic cathedrals—that searching is a multigenerational enterprise, and

The first antennas of the Allen Telescope Array, in California

no one should either expect early success or be discouraged by a few more decades of failure.

But that's being inconsistent. Drake and Sagan pioneered many of the ideas that underlie SETI. They motivated our hunt for extraterrestrials with their arguments. So it would be disingenuous to dismiss their estimates for the number of transmitting worlds while still accepting their rationale for searching.

Yes, it's possible that intelligence is truly rare. That's precisely what SETI is trying to disprove. Clearly, if the universe has been niggardly in populating its swarming worlds with human-level sentience, then SETI will remain an effort without triumph. But we don't know that yet, and given the large appeal and small cost of looking, the potential for endless failure shouldn't dissuade us from trying.

Another possibility is that aliens may exist but are deliberately cryptic, either because they have no curiosity about other societies or because they've decided that it's in their own interest to remain mum. But that suggestion bespeaks some insight into alien sociology, insight that we just don't have. We can theorize endlessly about the behavior of putative extraterrestrials, but doing so is like trying to guess the fantasies of your pet Pekingese; it's hard to know if you're right.

So while we can easily talk ourselves out of a search on the basis of such conjectures, doing so would be to surrender to mere opinion. In the 18th century, learned Europeans offered many reasons why a large, unknown continent might be located inside the Antarctic Circle. This argument undoubtedly provided good dinner conversation, but sailing south—doing the experiment—was the only way to replace speculation with fact.

However, there's a third possible reason why SETI might wreck on the rocky shores of failure: the myopia of its scientists. Could its practitioners be too provincial, clinging to the belief—as so many of them do—that extraterrestrial intelligence will seek contact via radio?

LOOKING FOR LASERS

Pioneering engineers like Nikola Tesla and Guglielmo Marconi were among the first to advocate radio for cosmic communication. It's an idea with a long run. But could this hoary history have biased us to think that tuning the microwave dial is the only promising approach to rooting out aliens?

Consider: The laser was invented in the mid-20th century, 70 years after Heinrich Hertz first produced radio waves on a lab bench. If that chronology had been reversed—if our local electronics outlet was Laser Shack rather than Radio Shack—would our SETI notions be different, too? Would we be building mammoth mirror-and-lens scopes for scanning the skies rather than welding together aluminum stampings for the Allen Telescope Array?

It's possible, but the SETI pioneers chose radio over light on the basis of efficiency. They claimed it took less energy per bit to send a message with radio. Their reasoning is easy to understand, at least if you passed high school physics. The smallest unit of either radio or light, called a photon, has an energy that increases as the wavelength decreases. A single photon of green light packs 500,000 times as much energy as a 21-centimeter radio photon, simply because its wavelength is a half million times shorter. That means that sending messages by radio ought to be a half million times cheaper than using light, at least as measured by your utility bill. For a conservation-minded Klingon, that could be important.

But in fact, the disparity isn't quite so imposing. Sending a single photon may not help Klingons to communicate if it becomes lost in the background noise. At radio frequencies a ubiquitous hiss fizzes through the cosmos—the leftover glow of the big bang. The fearsome fury of the universe's birth (now muffled by 14 billion years of cosmic expansion) is most bothersome at short radio wavelengths. A single radio photon, even at 21 centimeters, cannot stand out against this pervasive natural static. You need a hundred or so radio photons, bunched together, to reliably convey a bit of information. By comparison, you need only one photon for visible light. So the energy-per-bit disparity between the two schemes drops from a factor of about 500,000 to only a few thousand. That's a little less daunting, but it still strongly favors radio.

However, there's one more consideration, and this time the light brigade really strikes back. By using lasers rather than radio transmitters, chatty aliens can more accurately focus their emissions, ensuring more efficient use of the energy consumed by their transmitting setups. If you have $50 million (or its Klingon currency equivalent) to spend on a telescope, you can buy an optical mirror that's maybe 33 feet across. For the same amount, you can construct a radio reflector that's closer to 330 feet in size. (The surface of a radio dish doesn't have to be as smoothly polished as that of an optical mirror, which accounts for the difference in manufacturing cost.) If the aliens use one of those 33-foot mirrors to aim their messages, they'll be focusing the laser energy two billion times more tightly than they would with a 330-foot radio antenna! This enormous concentration of transmitter power—the incredibly small beams that can be achieved with inexpensive optical mirrors—more than makes up for the deficits suffered by light in its battle against radio.

Bottom line? The arguments for SETI experiments that would look for light pulses are as persuasive as those favoring a hunt for radio squeals. We should do both. And to some extent, we do. Both the SETI Institute and the University of California at Berkeley have optical SETI observing programs running in northern California. The scheme at both sites is similar. In particular, the eyeball end of each telescope is fitted with a few photomultiplier tubes. These specialized light detectors are responsive enough that even a single photon will set them off.*

**Several photomultipliers are used in these experiments to help rule out noise events that might occur in a single tube and that have nothing to do with flashing lights in the sky. Dan Werthimer at Berkeley has developed electronics to ensure that only flashes simultaneously seen by all tubes are recorded as a "hit." The telescopes are conventional, having mirrors that are roughly three feet across.*

The scopes are pointed at nearby stars and configured to nudge their astronomer masters if the photomultiplier tubes swallow more than a handful of photons in any nanosecond (billionth of a second) interval. Of course, they'll inevitably receive some light from the star itself. Our Sun, which is of mediocre luminosity, boils off about a billion trillion trillion trillion photons every second, as anyone who's counted them is aware. If ET's star does the same, then even at the relatively short distance of 100 light-years its torrent of photons has been reduced to a trickle, and during most nanosecond intervals, not even a single photon will hit the telescope mirror.

So if our SETI detectors suddenly announce the arrival of a few hundred or a few thousand photons within that short slice of time, well, someone's deliberately firing light pulses our way and you can start planning a new life, beginning with a check from the Nobel Prize committee.

The scheme for so-called optical SETI is very straightforward—you just have to build some hardware to look for very short, very bright flashes emanating from someone else's star system.

But how reasonable is it to assume that the aliens can build a laser capable of outshining their own sun by thousands of times? Answer: very reasonable. This is a claim I make with confidence because *we* can build such a thing. The most powerful pulsed lasers around today—if focused with a lens or mirror a few tens of feet across—could send an easily detected light flash to the stars.

People engaged in optical SETI have so far examined a few thousand stars without logging any convincing light bursts. At Harvard, longtime SETI practitioner Paul Horowitz has supervised the construction of a specialized telescope that scans the whole sky,* looking for brief flashes of light. The results to date are similarly drab: no bright pings. But optical SETI, while still receiving only a fraction of the attention that radio searches have garnered, is an inexpensive and growing enterprise. If we fail to find the aliens, it won't be because we haven't seen the light.

SEND THE PARTICLES

What of other communication schemes? Light and radio are not really different. They're just two spectral flavors of the same polysyllabic phenomenon: electromagnetic radiation. Maybe an alien society whose accomplishments trump ours would regard the idea of electromagnetic signaling as amusingly quaint. My correspondence suggests that this is an opinion held by some humans today:

"Why aren't you looking for gravity waves? Why would the aliens use radio, which only travels at the speed of light?"

"I believe I was contacted telepathically by a person from another star. When this happens, communication is by transfer of feeling."

"The aliens will use hyperdimensional physics, not archaic radio, to send messages."

"What about spooky action at a distance? Can't we communicate that way?"

"Dr. Shostak, learn how to uncover the vibratory level of

**You can see every star in the sky only if you live on the Equator or in space. In Massachusetts, the southernmost part of the celestial sphere is hidden from view. So the Harvard instrument can only examine about two-thirds of the full sky. In the Southern Hemisphere, the situation is reversed, and stars in the northern skies cannot be observed. You pays your rent and you takes your choice, although any astronomer will tell you that the southern skies are more spectacular.*

incoming energy and you will find endless visitors to this planet. Others have!"

Several of these suggestions invoke either physics we don't know, or schemes (such as telepathy) that don't seem to work in the laboratory (or, for that matter, at the gaming tables in Las Vegas). This isn't to say that aliens aren't blathering away via some sort of subspace communication system (à la *Star Trek*). But since we have no idea how such physics works (or for that matter, if any such physics really exists), we can't build the requisite equipment for a SETI search. For that reason, telling me to look for ET using highly advanced zonian vibrations, while well intentioned, is useless.

But some of the suggestions involve natural processes we *do* know about. A quick tally of my e-mails suggests that the alternative approach recommended most often is to use gravity waves, undulations in the geometry of space that were predicted by Einstein a century ago. Perhaps this is because large-scale gravity wave detectors, such as the Laser Interferometer Gravitational Wave Observatory, get a lot of column inches in the newspapers. But the big selling point for this approach is the belief that gravity wave communication will be a heck of a lot faster than radio or light beams because, in many people's minds, gravity is assumed to act instantaneously.

That would be a definite plus. After all, if nearby aliens are hundreds of light-years off, conversation by radio transmitter or laser will be draggy: You'll be filing your nails for a few centuries while a simple question bridges the distance to your chat mate, and you can do about a million crossword puzzles while awaiting the response. Presumably ET would prefer to phone home (or phone us) more quickly.

While a decrease in wait time is a worthy incentive, it doesn't seem likely that gravity waves would actually speed the pace of palaver. Yes, Isaac Newton assumed (with misgivings) that gravity operated instantaneously, but natural philosophy has moved on. Einstein's 1915 general theory of relativity predicted a more modest speed of interaction: Gravity waves cruise at light speed. While there are modern hypotheses for a different value (one reckoning insists that these waves travel 20 billion times *faster* than light), a compelling experimental verification for one value or the other has yet to be made. The smart money (which is a euphemism for "most physicists") figures that conversing on a gravity wave would not be any snappier than light or radio. In addition, building a usable gravity wave transmitter is a mammoth undertaking. Transmitters able to generate powerful radio signals do so by shaking electrons. The requisite hardware fits in a two-car garage. To transmit an easily detectable *gravity* wave demands engineering on a somewhat weightier scale: You need to vigorously shake something the size of a planet or star. It's hard to see the appeal.

Neutrinos have also been put forth as an attractive alternative to our usual SETI approach. Perhaps this is because these high-speed, high-energy particles are so staunchly relentless. They easily pass through the Earth and just about anything else. As you read this sentence, a hundred trillion neutrinos—most of which originated deep within the Sun—have hurried through your head. The distraction is small.

Indeed, there are proposals to take advantage of neutrinos' unstoppable nature by enlisting them in the service of global communication. We could dispense with undersea cables and orbiting satellites, and send messages halfway around the world by aiming a neutrino transmitter on a shortcut through the earth. The military could use directly beamed signals to communicate with its submarines, without fear of jamming.

Beyond that, if neutrinos are truly "silver bullets" that can slice their way through such interstellar impediments as gas, dust, and an occasional planet (a handy talent if your own world has rotated so that you're not facing the alien transmitters when their message arrives), wouldn't ET choose them as the preferred way to communicate?

Possibly, but alien neutrino messages might get lost in the noise of neutrinos produced by nature. The problem can be avoided by using neutrinos of very high energy, since highly energetic particles would easily stand out from the madding crowd, but the downside is that they're terribly expensive to produce. For example, one proposal has suggested that aliens might signal with neutrinos having an energy of 6,000 trillion electron volts. Electron volts are units of energy that might befuddle the nontechnical, but suffice it to say that a million such messenger neutrinos would carry the same wallop as a .45-caliber slug. That may not disquiet you, but it should. To get our attention, an extraterrestrial society would have to shoot billions of billions of such neutrinos our way.

Since cost of communication is always an issue (consider your cell-phone service), let's work out the energy expense of neutrino signaling. Imagine that in their attempt to ring our bell, the conservation-minded extraterrestrials shower neutrinos only on that fraction of our solar system lying between the orbit of Venus and Mars. They do this on the advice of their generously paid and highly respected astronomers, who claim that the best planets for life would be situated in that narrow neighborhood. Let's also assume that, for our part, we've built a neutrino detector more than 0.2 cubic mile

(1 cubic kilometer) in size,* able to notice one particle in a thousand (a pretty good hit rate: Most of today's neutrino experiments detect one particle in a million). To attract our attention with only ten detected neutrinos—an uncomplicated message that says no more than "we're trying to get in touch"—will require as much energy as is burned every day by all the cars, trucks, trains, boats, and planes on Earth. If the aliens, in their fervor to find new friends, are pinging likely worlds at the rate of one a second, then their energy bill will be 300 million million million dollars a month, assuming they pay the same utility rates you do. Even for well-heeled aliens, that might be burdensome. In sum, and despite their popularity, neutrinos are a gold-plated communication scheme.

Another suggested scenario for interstellar communication is the equivalent of snail mail. In the 1990s, when Project Phoenix was deployed at the big radio telescope at Arecibo, we used an Internet connection that could send data at the rate of about 100,000 bits per second back to the SETI Institute. That's speedy enough to transmit the text of this book from the Caribbean to California in about a half minute. But consider another data transport mode: the U.S. Postal Service. Instead of pumping those data through the Internet, we might simply have written them to DVDs and mailed them. Suppose we collected 100 such DVDs, boxed them up and dropped them at the post office. Three days after doing so, four trillion bits of data would be delivered to the institute's

**A neutrino detector of comparable size is now being built by the University of Wisconsin. Situated under the ice in Antarctica, it will be used to look for highly energetic, natural cosmic events far beyond our galaxy, such as exploding stars, black holes, and gamma-ray bursts. IceCube, as it's called, will be one cubic kilometer in size.*

doorstep. That works out to an average transfer rate of 15 million bits per second, or 150 times faster than our Internet connection.

Christopher Rose and Gregory Wright, computer scientists at Rutgers University, took this idea to what they think is a logical conclusion. They argued that we shouldn't expect ET to prattle into a microphone. Instead, they figured it more likely that the aliens will write their messages onto disks, thumb drives, flash cards, or their ET equivalents, stuff these digital greeting cards into interstellar rockets, and fire the missiles off toward their extraterrestrial pen pals. The computer scientists' claim is that going postal could be far more frugal than broadcasting radio signals to someone else's solar system, requiring as little as a trillionth as much energy for the same message.

Does this mean that we should be looking for "messages in bottles" rather than signals? Could an advanced civilization have seeded our solar system with a packaged dispatch we have yet to find?

A few sticky practical difficulties could keep the aliens from flooding promising star systems with epistles in missiles. The fundamental problem is that rockets are slow, and you need a separate one for each delivery (unlike radio, which can broadcast to many stars simultaneously). Even a very fast rocket is likely to take many millennia to sail from ET's world to ours. Since the extraterrestrials didn't know of our existence thousands of years ago (after all, no radar or TV signals would have reached their planet), they'd have had little incentive to launch a rocket busting with bits to our solar system. From the aliens' point of view, this would be very expensive junk mail. In addition, any extraterrestrials clever enough to find Earth and its lacquer of life would have enough technology to carefully target their radio or laser broadcasts, thus greatly reducing the energy costs. These considerations suggest that, while some sort of interstellar postal delivery might indeed be set up by communicative aliens, signaling seems more practical for making first contact.

Another alternative to conventional SETI has been worked out by Walt Simmons, a physicist at the University of Hawaii. He and his colleague Sandip Pakvasa devised a clever scheme that would allow interstellar broadcasters to keep the coordinates of their home planet secret. The trick is to forgo conventional electromagnetic signals made up of large, organized "waves" of photons in favor of individual, quantum-entangled photons.

The idea entails more than merely substituting a small task force for a large army. Individual photons can be quantum-mechanically related—they can have buddies, if you will, with which they share information. Each buddy is sent separately by the broadcaster and reunited with his pal at the receiving end. They deliver their message only when they're brought together.

In practical terms, the way this might be accomplished is that each member of a photon pair is sent in opposite directions from the broadcaster's home planet. One might be beamed a light-year to the left, and the other a light-year to the right. They would be aimed at mirrors that would redirect them to the target star system. Additional nonpaired photons could be sent along as well, to swamp the presence of these message bearers, somewhat like using disorganized street crowds to hide a riot squad.

At the receiving end, the photons that came from one mirror or the other would be indistinguishable from cosmic background noise. But the quantum-entangled buddy photons would unite to form a microscopic image—a picture. The picture, of course, could contain all sorts of interesting information that sophisticated aliens might wish to share with us or with others in the galaxy.

As noted, the image would be quite small. And looking at it would disturb it in a way that would bring Heisenberg's famous uncertainty principle into play. In fact, by reading the message, all the information about its origin would be lost. This would be more closely akin to communicating via bottled messages than the postal system described earlier. The message would arrive, but the sender would keep his location secret.

According to Simmons, "It's fairly straightforward to target specific recipients, and it wouldn't be hard for the transmitting society to methodically send messages to large numbers of star systems, one after the other."

Could quantum messaging dominate interstellar communication? Could this be the preferred way to get in touch with unknown cosmic beings? If so, it offers an appealing resolution of the Fermi paradox, which asks, "If the galaxy is teeming with intelligence, why don't we see evidence for it everywhere?" Perhaps the evidence *is* everywhere—washing over us right now in a shower of quantum-encrypted messages.

Sadly, we don't have the technology to look for such signals today, although Simmons expects that the requisite instruments are less than a century away.

"Meanwhile, we should continue our SETI searches," he advises, "we should absolutely do that." After all, there are many ways to get in touch. It's just that some of them don't carry a return address.

I've described alternatives to the usual SETI approaches at some length, partly because of the popularity of such "different strokes." Some people spend endless hours crafting e-mails addressed to me and my colleagues wherein they point out all the reasons we are wasting our time looking for radio signals or light pulses from aliens. "So old school," many of these critics opine, and as we've noted, other schemes certainly can be considered. Several alternatives require technology (if not science) that's still over the horizon, so there's little point in suggesting that SETI should switch strategies. You might as well have told Orville Wright to use jet engines or, more aptly, suggested that James Cook await turbine-powered ships.

Not only can't we effect such a dicey switch in paradigm, it's not clear that doing so would be a good idea. Exploration is always carried out with the technology of the time—and discoveries occur not because that technology is optimal for the task at hand but because it's just good enough to succeed.

We also are not ruling out a well-known phenomenon in science loosely termed serendipitous discovery. A famous instance occurred in 1896, when French physicist Antoine-Henri Becquerel happened to put uranium crystals into a darkened drawer with some photographic plates. He unexpectedly found that the latter were exposed by the former. This accidental revelation of natural radioactivity prompted the study of nuclear physics. So even though SETI researchers are not building devices to detect gravity waves or neutrinos, there's always the possibility of a serendipitous discovery of extraterrestrials by scientists who *are* constructing such devices.

In any case, I want to emphasize that it's highly premature to be pointing to SETI's lack of success and blaming it on a faulty strategy. We have carefully examined only 0.0000005 percent of a single galaxy.

But that percentage is going to climb significantly in the next several decades. What if we did find a signal? How would we know, and what would the public think?

CHAPTER FIVE

THE SMELL OF THE OIL

To outsiders, sitting around waiting for ET's signal seems about as exciting as Uncle Sid's calliope music collection. That's because outsiders judge SETI on the basis of what they've seen in the movies, and the movies get it wrong.

First off, culling the band for signals using those curly audio transducers known as ears is a nonstarter. Sure, you could convert the incoming radio noise to sound—any radio can do this—but when you have 100 million channels, which ones do you pick for monitoring? The question is moot for machinery: Computers can comb through every one of those many millions, and do so without boredom or bathroom breaks.

Since the actual signal monitoring has been outsourced to digital electronics, why do researchers hang around the radio telescopes at all? For many experiments they don't, and that situation will eventually extend to *all* such efforts. The scientists will stay shuttered in their offices, usually hundreds of miles from the antennas, dully sifting through e-mail and worrying about the next experiment. They can be confident that the software will notify them if something interesting has been found by the telescope's ceaseless scrutiny of the heavens.

SETI is being automated, so the tedium will be eliminated. But some of the romance will go too. That romance derives from more than just the idea of looking for aliens. It's palpable, because there's no experience like being at the telescope.

For a half dozen years, I made the twice-yearly trek to the Arecibo radio telescope to take part in Project Phoenix, our perusal of about 750 nearby star systems. We used the antenna at night (during daytime the Sun might fatally scramble the narrow-band radio signals we look for), and that meant we would split the observing into two six-hour shifts hinged at midnight. Typically, I rode herd on the telescope in the evening, and Jill Tarter would take the graveyard shift. Jill said she preferred this late-hour observing because the data were somehow more exciting then.

Jill's statement may speak to the lack of interference after midnight. Then again, it might just be a perception induced by the wee hours, or possibly Jill's penchant for playing salsa music in the observing room at a volume loud enough to kill plants. But even in the supposedly calmer, earlier hours of the evening, we received signals all the time—a dozen a minute was commonplace. Most of these signals were immediately recognized by our software as man-made interference. Some signals took a little longer to figure out: about ten minutes or so. Once every few days we'd get a "hit" that looked good for a half hour or more. Observing would be pretty dull without an occasional adrenaline shot.

Being at the radio observatory was like attending summer camp. We worked there, ate there, and slept on site. Diversions and entertainment were minimal. Once every week or two, we'd organize an expedition into the town of Arecibo itself, about a half hour drive away, to visit a chain restaurant for a burger. We were like sailors on shore leave, although our activities were less indecorous.

Although routine, observatory life was pleasant and mildly addictive. Every so often, I'd step away from the logbook and the computer displays, abandon the chitchat with the engineers, and walk outside into the humid dark of a tropical evening. At the bottom of a precipitous slope was the antenna, hunkered down in the knobby Puerto Rican karst, with its unblinking aluminum reflector curved toward the sky. No sounds of traffic broke the quiet, no distant squeal of kids playing, not even the raspy roar of jets overhead: only the gentle, high-pitched chorus of small frogs—the *coquis*—and the slow, baritone grind of the telescope's tracking motors. A few lights sparkled from the focus platform, but the forest was otherwise murky and dank. The air was tinged with the smell of lubricating oils—the common scent of all radio observatories.

I've sometimes stood there for an hour or more, as the telescope's metal ear faced off the stars one by one. This was the dull monotony of exploration, as machinery the size of a city block sought, and sought again, a delicate dance of electrons choreographed by beings on worlds we have never seen.

EUREKA

It was just past 11 p.m., and although I wasn't tired, I nearly swerved into the K-rail separating Highway 101's north and south lanes. This near loss of control was caused by a late-night talk program, *Coast to Coast AM.*

Coast to Coast is a hugely successful national radio show begun by Art Bell. His broadcast career first bloomed at a local Las Vegas station during the early 1990s, and he soon leveraged his success into a nationwide call-in program. He routinely drew an audience of more than ten million listeners on more than 400 AM stations. Bell eschewed the usual talk-show topics: politics, home finance, or how to configure your laptop. Instead, he focused on less worldly concerns, chatting with guests about such frothy esoterica as UFOs, out-of-body experiences, remote viewing, and the impending end of the world.

It was the fall of 1998, and Bell was churning a big story—what he claimed was a "revolutionary discovery." Only a few days earlier, on October 22, humans had apparently made contact with an alien civilization. An Englishman had hacked into the website of an amateur SETI group known as the SETI League to bare the startling news.* The hacker posted the following message:

"Hello, I am sending this along to inform you of a possible SETI hit. . . . I am an engineer at a major telecommunications firm in England and I have for the last year conducted SETI from one of the large (10-meter) dishes we have here that was taken off-line some years ago. This is a 'parasitic' experiment and no one around here knows I mounted a second feed horn on the dish due to the fact that I am the engineer."

The hacker went on to describe his equipment and then to announce that after tuning to 1453 MHz, he had picked up a signal coming from the nearby star system EQ Pegasi. He refused to sign his name, implying that to do so would be personally dangerous.

This story, which might cause your eyebrows to head north for several reasons, was quickly picked up by the BBC's online news service. On October 29, that august organization carried an article declaring that "the scientific world is buzzing with the suggestion that signals from aliens . . . may have been picked up by a part-time astronomer."

This tentative endorsement by the world's largest broadcasting corporation galvanized the talk-show host. As Bell continued to speak of this stunningly important development, he kept saying "and I hope that Seth calls in to the show."

Turning on the car radio to hear this sort of fare was like finding a horse head in your bed.

**The SETI League, largely the brainchild of New Jersey electrical engineer Paul Shuch, recruits amateurs to use backyard satellite dishes and homebrew electronics to search for signals. Several hundred volunteers take part.*

Twenty minutes later, when I arrived home, my wife, Karen, announced, "Art Bell wants you to call him. He left his number." Moments later, I was on the air. Bell was enthused about the claimed detection and wanted my opinion.

Well, I had one, and it was decidedly skeptical. The first thing that made me suspicious was the name of the star system: EQ Pegasi. Roughly 21 light-years away, EQ Peg comprises two runty stars of the type known as M dwarfs. Only a month earlier, on the night of September 15, Jill Tarter and I had been at the Arecibo telescope observing for Project Phoenix. When this stellar twin was in our sights, our receivers picked up an intriguing signal. For roughly ten minutes, it was looking good: It seemed to be extraterrestrial. However, further tests showed that the signal was just more radio junk from Earth. More specifically, it had the hallmarks of a satellite in low Earth orbit.

This minor episode was not uncommon, and I wrote it up for the SETI Institute's website and also as part of a series of articles I was penning for MSNBC. I figured that the EQ Peg false alarm was instructive in showing how we sort wheat from chaff.

So when, a month later, some anonymous "amateur" claimed to find a signal coming from EQ Peg, I was immediately suspicious. How coincidental! After all, there are 88 constellations in the sky. So there's about a one percent chance that a real signal would appear in the same star field as the interference signal I publicly described only weeks earlier. The fact that the signal was emitted by the identical dim star was even less probable.

Then there was the disquieting fact that the announcement of the signal—a discovery that would rank among the greatest of the last century, if not the last millennium—was made *anonymously* and via hacking someone's Internet chat group. That procedure didn't have the aroma of real science.

I told Art Bell all this. I also told it to him again a few nights later, when he invited me back on the air in the wake of new Web pronouncements claiming more detections of the EQ Peg signal by amateurs in Guernsey and Japan. The matter still reeked of a hoax. The evidence presented for any of these signals was poor to begin with, and never improved.

As a single example, the original detection—the one first described on the SETI League's website—was said to be a so-called drift scan. In other words, the English engineer was suggesting that his borrowed antenna was fixed in place—its reflector didn't move. Consequently, EQ Peg would slide into and out of its beam as the Earth turned, causing the signal strength first to wax and then wane. However, the graph of the signal as actually posted showed no change in intensity, as a true drift scan would. When this fact was pointed out, new scans were suddenly proffered that accommodated this small technical requirement. It's always convenient when one can change the observations to fit the discovery.

Only a few days later, the claimant reported on the Internet that measurements with the 300-foot Effelsberg Radio Telescope, near Bonn, Germany, had confirmed his result. However, when I contacted the folks at Bonn, I learned that they had never looked at EQ Peg. The Englishman's claim was a simple lie.

Shortly thereafter, the anonymous poster wrote that he would hold a press conference on November 4 in London, under the auspices of the International Astronomical Union and the British Interplanetary Society. This sounded reputable, and proved to be the tipping point for several members of the research community. After all, if the British Interplanetary Society was taking this seriously, perhaps they should as well. Rather than continuing to whinge about inaccuracies and inconsistencies in the EQ Peg claims, a few academics decided that they had no choice but to go to their radio telescopes and check out the signal firsthand.*

On November 2, Australian radio astronomer Ray Norris used the Australia Telescope Compact Array to observe EQ Peg. The ATCA, consisting of six antennas each 72 feet in size, has 30 times the sensitivity of the dish the British amateur claimed to be fielding. But Norris didn't find the signal. John Whiteoak, an Australian colleague of Norris's, tried to track down the wily EQ Peg aliens using an antenna at Mopra, also in New South Wales. He didn't find a signal either. In Massachusetts, drift scans made by Paul Horowitz's group at Harvard similarly came up empty, as did attempts by a half dozen members of the SETI League.

For anyone who thought this coffin needed more nails, the highly touted London press conference also turned out to be fictive. Shortly thereafter, all the Web postings relating to the claimed EQ Peg detection evaporated. They were replaced by a menacing U.S. National Security Agency emblem, presumably to imply the nefarious involvement by the American government.

At this point, Richard Hoagland (he of the Face on Mars) became Art Bell's expert on the story. Unlike me, Hoagland wasn't chanting the tiresome tune of "this is a hoax." He not only claimed that the EQ Peg signal was real but further stated that the signals were coming from an alien probe, inbound and headed for a landing on Earth. Hoagland vaguely

---

**The initial skepticism of SETI researchers provoked some members of the public to accuse the scientists of sour grapes—the "professionals" were cheesed off because the signal had been found by an amateur.*

referred to proprietary inside sources at the Pentagon who told him that whatever was transmitting the EQ Peg signal was going to alight on our planet on December 7. He also told Bell's audience that this epic touchdown would take place somewhere north of Phoenix, Arizona.

Stunning insider knowledge. But as far as I know, December 7, 1998, came and went without any aliens wandering into the desert north of Phoenix asking for directions or a cold beer. Nonetheless, Hoagland continued to talk about the EQ Peg signal well into December, when the story completely lost its legs.

EQ Peg was more than a false alarm. It was a straightforward prank. The phony claim wasn't notably ingenious, nor was it impressively sophisticated. Pulling it off required only the chutzpah to manipulate some readily available SETI League graphics, pour them into a Web page, and garnish the whole thing with pretension and bad grammar.

When this less-than-genial practical joke ended, SETI researchers were understandably chagrined. Several of them had been seduced by trickery to the point of burning up expensive telescope time (not to mention their own man-hours) to confirm the signal. If an unpolished Internet fiction was good enough to bamboozle SETI astronomers, you might wonder whether academics would be clever enough to handle the real deal. If a broadcast from ET were to wash over their high-tech antennas, would these SETI types even recognize it? Conversely, if these same folks were to appear on the nightly news, breathlessly informing us that they'd found the aliens, could the public believe *that?*

To answer these questions, we need to grasp exactly what SETI experiments are looking for. What type of signal would convince the experimenters that ET was really on the air? And if SETI researchers *were* convinced, how would the news be released; what's the "protocol" for letting everyone know that we've downloaded a broadcast from an alien society? Clearly, while many people are aware of SETI's work, they are less sure what would happen in case of a success. And many suspect that the scientists wouldn't be entirely forthcoming about their discovery.

HOW WE KNOW IT'S ET

One question I'm never surprised to get (mainly because I get it more often than I get haircuts) is whether extraterrestrial correspondents would preface their communications with a string of prime numbers, or maybe the value of pi. Such an introduction would give us a vital clue that the signal is deliberate and nonnatural. This idea made it into the movie *Contact,* after all, and if it was worthy of Carl Sagan, wouldn't it also be an obvious strategy for ET?

Without doubt, if we picked up a series of radio pulses drumming out mathematical bons mots, we'd pronounce the signal "artificial." Giant clouds of interstellar gas, which fill the airwaves with their own cacophonous broadcasts, don't seem to know much about pi or prime numbers. On the other hand, any society able to build a powerful transmitter would.

We should start at the beginning, though. Most of our SETI experiments are oblivious of message content. That's because we deliberately average the radio noise picked up by the telescopes for seconds or minutes at a time, thereby helping to make the signal stand out from background static. Regrettably, this averaging obliterates messages. A television signal averaged for even a millionth of a second is denuded of its picture and sound, the parts that many people find most interesting. The effect is similar to listening to a concert in a room with heavy reverberation—all the piccolo arpeggios are smeared out and lost. So when we average incoming cosmic static, you can be sure that any alien message has inadvertently been tossed out like a bathed baby. But we average anyway, because that allows us to find weak signals. Our first priority, after all, is simply to learn that the big red "Alien on the Air" sign is illuminated. Only later will we worry about finding out what ET is mumbling into the microphone.

This fact—that we're looking for the radio equivalent of bottles washed up on the beach, not the crumpled bits of paper that might be inside—is little appreciated by the public. Maybe that's a good thing, because I suspect they would be dismayed to hear that we won't be able to immediately access ET's message, unlike in the movies, when attempts to decode the incoming radio traffic usually begin right away. Finding the message would come later, possibly quite a *lot* later. As a sobering example, I note that extracting the picture and sound from a television broadcast requires roughly 10,000 times as much signal sensitivity as that needed simply to detect the transmitter. If alien broadcasts are similarly encoded, then we'll undoubtedly need to build an antenna many miles across to snag the information contained in an extraterrestrial ping.

My colleagues aren't fazed by this obstacle. They assume that if we discover a broadcasting society with today's modest SETI instruments, the money for the giant telescope needed to find a message would explode from government coffers. That sounds right to me, but it's possible we're all being naive. The discovery of the Dead Sea Scrolls didn't prompt enormous cash outlays to archaeologists so they could decipher the meaning of these tattered texts. Ditto for such mysteries as the Rongorongo script of Easter Island or even Stonehenge.

But those cryptic messages were left by societies whose level of knowledge was far lower than our own. Yes, decoding Stonehenge would provide interesting information about the burial practices and daily drudgery of Druids and their predecessors, but that's about all. An alien message would be emphatically different, as it might contain information far beyond what we know, information that could transform our future. So I suspect that on the heels of a detection, a fat government check would be in the mail, and SETI would morph into big science.

Still, if it's not the message that proves a signal to be of alien origin, then what does? The answer is simple: location, location, location. Where is the signal coming from, both now and a few minutes from now? A trivial example of how this works is a transmission whose apparent location is always the same. Such a broadcast is obviously due to a transmitter bolted to earth. The sky, after all, whirls above our head. A stationary signal might be radar from the local airport or some other earthly emission bouncing around our telescopes, but it's clearly not a Klingon "hello."*

Unfortunately, merely confirming that a signal source changes position doesn't guarantee its extraterrestrial pedigree. We have to show that it moves the way the stars do, spinning around the Earth roughly once every 24 hours.

This task is greatly complicated by the swarm of artificial satellites that sail the skies. This phalanx of orbiting birds, many of which have been lofted into nearby space

**According to Internet dictionaries, "hello" in Klingon is* nuqneH *(pronounced nook-neck).*

to accommodate our need to talk to distant relatives, has become a sizable disturbance: Nearly a thousand satellites are beaming transmissions to Earth, producing signals of the type that our SETI experiments seek. This causes endless confusion and consternation.

So how do we separate ET from AT&T? Fortunately, the majority of this high-altitude hardware is in motion against the background pattern of the stars. Most telecommunication satellites circle the Earth every 90 minutes or so, whirligig behavior that makes them easy to identify. Other satellites (the ones that feed the satellite dish on your balcony with an unwholesome number of TV channels) are in so-called geosynchronous orbit. The geosynchs orbit Earth at a speed that keeps them apparently fixed in one spot on the sky, a great advantage when setting up your TV dish. But that sets them apart from the stars as well. Indeed, they don't seem to change position at all, a situation that's unlike any celestial object other than Polaris, the North Star.

Consequently, determining if a signal is extraterrestrial is mostly a matter of pegging its position and measuring its motion. We employ both simple and not-so-simple methods to do this. One approach is to use two radio telescopes, separated by hundreds or thousands of miles. We used this scheme throughout the nine years of Project Phoenix. If a signal was picked up by one telescope—say the hulking Arecibo antenna—we would check it immediately with a second dish, in this case the 250-foot Lovell radio telescope at Jodrell Bank, England. I say "check" as if we actually rang up England. In fact, the checking was done automatically by computers. Note that if we found a truly promising candidate signal, we'd seek independent confirmation from an observatory not involved with our own project.

Naturally, both of the telescopes would detect a broadcast from a distant planet. But if the radio noise were just the burping beeps of a communications satellite gliding over the Caribbean, then Arecibo would see it but Jodrell Bank wouldn't. We'd know right away that we could reject such a signal.

Even if we had the bad luck to be confused by a satellite high over the Atlantic that was detected by both antennas, our two-fisted strategy could still sort out the location of the transmitter based on the measured frequency of the arriving signal at each of the telescopes. The frequencies would be different for a celestial source than for a man-made transmitter pirouetting around the Earth at a couple tens of thousands of miles an hour.

In the course of Project Phoenix, searching with two antennas greatly improved the efficiency, and therefore the speed, of our search. But the use of the British dish cost the privately funded SETI Institute significant money—about $3,500 a night, plus pub meals for our engineers.

Another, less expensive scheme for nailing down the position of a signal source is simply to look now, and then look again later. As an instructive example, Dan Werthimer and his colleagues at the University of California at Berkeley run a radio search on the Arecibo telescope known as SERENDIP (the acronym stands for Search for Extraterrestrial Radio Emissions from Nearby Developed Intelligent Populations, a small fact that was probably obvious).

SERENDIP is a so-called commensal experiment. Now, that may sound like a 19th-century enclave where people share rural labor, the word of God, and each other's husbands and wives, but in fact it's a SETI program that piggybacks on a telescope already being used for other projects. In the case of the Berkeley experiment, the researchers have commandeered an unused receiver on the Arecibo antenna that swings without direction around the sky, pointing at who knows what.

The scientists collect a continuous stream of radio noise from this peripatetic instrument, hoping that by chance the array will occasionally be aimed in the direction of a transmitting world. If a signal is found (and given all the terrestrial interference, this is not a rare event), it's duly logged by computers. If this same signal is found a second and third time (usually many months or even years later) at the same spot on the radio dial and at the *same position among the stars,* then it becomes a serious candidate for being ET and will be checked out specifically by the scientists.

Since there's no rush in analyzing the data from their commensal experiment, the Berkeley researchers farm a small fraction of it out for others to process. In particular, some SERENDIP data are made available on the Internet for examination with the popular (and free) SETI@home screen saver, a project in which people around the world donate number-crunching time on their personal computers.

The idea is simple: Once you've installed the screen saver, it will download a packet of data from the SERENDIP experiment. If you walk away from your computer to make a cup of coffee, the software begins combing through the data packet looking for the type of narrow-band signals that an extraterrestrial transmitter might emit. If it finds such a signal, its characteristics are reported back to Berkeley, along with information on who processed the packet. The same data are distributed to multiple processors, incidentally, to foil those who might wish to fake a detection.

Currently, the number of people who have downloaded SETI@home amounts to more than eight million. Any one of them could, in principle, attain eternal fame, even if not much fortune.

A third technique for pinpointing a radio source is to "nod" the telescope. Suppose a possibly extraterrestrial signal is received. The computers can check it out by simply tipping the antennas a few degrees away from the target to see if the signal goes away. Interference from man-made sources is usually so strong that it doesn't much matter where the antennas are focused: We get a signal everywhere. This interference is akin to the light from a desk lamp turned on in a dark room—you see the illumination no matter where you look. But ET's transmitter would be a tiny lamp on the ceiling somewhere. Since the source is much weaker, we won't measure its emission if we're not accurately pointed at the right position.

This "moving off-source" scheme is trustworthy to the point of being bulletproof. It provided the first indication that the signal of June 23, 1997, came from the SOHO satellite (see page 18). Nodding, however, while reliable, is inefficient—think of a moose hunt where you stop to do DNA tests on every likely animal to be sure it's a moose. You'd spend all your time making tests rather than scouting for ungulates. Fortunately, the new Allen Telescope Array can do the equivalent of this nodding in a simple and elegant way. Unlike nearly all of its predecessors, the array—like the mythological giant Argus Panoptes—has multiple "eyes" and can observe more than one star system at a time. Imagine simultaneously examining two stellar positions. One can be used as the check position for the other. If a signal is found in both directions (and at the same radio frequency), then it's interference. If not, it might net you a ticket to Stockholm, a lot of open-face fish sandwiches, and the Nobel Prize.

The Berkeley SETI group has recently added a new wrinkle to its SERENDIP program: Astropulse. Home computer users can now process the same data distributed for SETI@home, but in a different (albeit more time-consuming) way. The Astropulse software looks not for narrow-band whines but for quick radio flashes, less than a millionth of a second long. The motivation for Astropulse is straightforward: Perhaps ET is trying to get our attention not with an endless "flute tone" but with an intermittent "snap." Indeed, thinking about the whole SETI problem from ET's point of view suggests that looking for short radio pulses might be a logical strategy for finding extrastellar life.

SETI STRATEGIES

Maybe because Frank Drake is one of our staff scientists, the SETI Institute has always favored the idea of searching for signals coming from solar systems that are cousins to our own. That strategy underpinned Project Ozma. The principle still sounds good, but our astronomical knowledge is limited, and we haven't yet discovered any such brethren solar systems. We believe habitable worlds abound out there, but we still don't know. Ergo, we direct our antennas toward stars that *could* have planets in the modest hope that some of these actually host Earthlike worlds, outfitted with garrulous creatures who are beaming signals our way.

We'll continue to pick out stars that could contain suitable solar systems, at least until we learn more about the galaxy's planet inventory and can better choose our targets. But a "serial" search for ET—one in which we look first here and then there—does have a dead-obvious difficulty. We don't spend very much time searching in any given direction.

That restriction has not escaped general notice. Hardly a week passes without some keyboard correspondent telling me that SETI has a honking big problem. "Even if the galaxy is brimming with intelligent life," they'll gravely intone, "your experiment has virtually no chance of success."

According to their reasoning, the probability that SETI's receiving antennas would be pointed at the aliens' world at the right moment to pick up a signal emitted when their transmitting antennas were aimed at our world is vanishingly small. Since we lack a *TV Guide* for extraterrestrial broadcasts, we'll never twist our telescopes in the right direction, at the right time, to pick up a signal. This has been likened to two bullets colliding by chance.

I'm not talking about a failure to connect because their civilization bloomed and died a billion years ago, or a billion years from now. The disappearance of technical societies, while a bummer if it's *your* society, is hardly an issue for SETI if the galaxy spawns them on a regular basis. There will always be someone out there, firing up their radio transmitters. Rather, this is a synchronization problem. SETI researchers typically view any particular star system, at a given frequency, for only a few minutes at most. But what are the chances that an alien signal will splash upon our antennas during that brief interval?

Well, the chances are 100 percent if the extraterrestrials are semi-immortal and beam interminable broadcasts to the whole galaxy (or at least a big chunk of it). If their program lineup is "All Alien, All the Time," then it doesn't matter *when* we look, as long as we eventually observe in the right direction at the right spot on the dial.

The problem with this scenario is that an omnidirectional radio broadcast really combusts the kilowatts. Consider an extraterrestrial transmitter halfway across the Milky Way, belching out the Galactic Headline News. Further suppose that their transmitter is beefy enough to produce a signal that is detectable by Project Phoenix. Phoenix, as noted, was the most sensitive targeted SETI search ever. The required transmitter power would be 100,000 trillion watts, or about 10,000 times the total energy consumption of *Homo sapiens* today.

That sounds like the mother of all utility bills, at least for the types of societies we can easily envision. On the other hand, it's only one-billionth the output of the Sun. We could corral that level of energy by, for instance, carpeting Mercury wall to wall with high-efficiency solar cells. Sure, we'd sacrifice a lot of airless Mercurian real estate (would you really care?), but we'd have enough juice to mount a pangalactic broadcasting project. And if *we* could collect such massive amounts of energy, maybe *they've* already done so. It's not inconceivable that some truly advanced civilizations are willing and able to field high-powered, omnidirectional beacons.

But there's another approach to transmitting, a "two tier" strategy that I think would make sense for prudent extraterrestrials. This strategy assumes that any aliens wanting to get in touch will be at least a century or two ahead of us. If so, you can bet your boots they've already built space telescopes that can find and analyze planets around other stars. If life is commonplace (still a big "if"), then the appendixes of these aliens' astronomy textbooks must be stuffed with long lists of worlds known to have biology. And we'd be on those lists. After all, for two billion years the oxygen in Earth's atmosphere—largely produced by photosynthesis—has been whispering to the galaxy that life exists on our planet. Without doubt, Earth would be an entry in those textbook lists of "bio-worlds."

Consequently, any sensible alien signaling project would target its transmissions to bio-world-blessed star systems. If the extraterrestrials were on a fishing expedition, why wouldn't they concentrate their efforts on those places where the fish are most likely to be? But we're still faced with that irksome synchronicity problem. Even though Earth has had detectable biology for two billion years, it has boasted technologically sophisticated life for only a century or so. Let's assume that our civilization lasts for another 20,000 years.* Then, doing the simple math, you'll realize that a random transmission to Earth would have roughly a 0.00001 chance of arriving when its inhabitants possess the technology for noticing a signal.

Putting this another way, if you're the program manager at a Klingon radio station busily targeting planets known to have life, you'd better ping at least 100,000 of them to garner even *one* world with intelligent listeners.

So here's the proposed strategy. The aliens decide to use a highly directional antenna to beam a short, intense pulse to a long list of bio-worlds. Let's say they repeat this serial signaling project every 20 minutes. Their instrument for doing so might be a souped-up equivalent of the Allen Telescope Array, but outfitted with transmitters rather than receivers. With certain design modifications that we can envision but still can't afford to build, an array like this could very precisely target star systems.

If each pulse is one-thousandth of a second long, their list could comprise about a million targets, enough to provide a

---

**Not everyone is so sanguine about our future. Despite the widespread belief in our own enlightenment, we might well self-destruct in far less than 20,000 years. When Mahatma Gandhi was asked by a reporter what he thought of Western civilization, he replied, "I think it would be a good idea."*

decent chance for finding a few star systems sporting creatures able to build big radio telescopes. That's the first tier of the strategy, devising a powerful beacon whose only purpose is to attract someone's attention.

The second step is even simpler. If we find that hailing signal (and it could be easily verified, as it repeats every 20 minutes in our example), you can bet that we'd pull out all the stops to examine the patch of sky from which it came, looking for additional signals. We'd reason that no one would send us an empty string of pulses; no one would shout to get our attention without wanting to say more. There must be a message somewhere. So the second tier of the fishing strategy, from the aliens' point of view, is to have a relatively low-power (and therefore economical) omnidirectional transmitter that carries the message, on the assumption that we would eventually build a big enough receiving setup to find it.

If any extraterrestrials are trying to get our attention with such a two-tiered scheme, we need to modify our traditional SETI observations to find them. In particular, we need to look for those short, intense hailing signals.

Berkeley's Astropulse project is already searching for radio burps less than a millionth of a second in length, and optical SETI experiments are sensitive to bright sparkles a thousand times shorter. But neither stares very long at any particular target, and they seldom come back for a second look. This failure to stake out a target for hours or days virtually guarantees that we will fail to find hailing signals that repeat only intermittently.

Now you might think that other telescopes—perhaps even the Hubble Space Telescope—might make up for this deficiency, since these other devices are frequently trained on one patch of the sky for long periods of time. Alas, their imaging devices, which are similar to those you'd find in a digital camera, cannot detect light pulses shorter than about 0.1 second. They could easily miss very short light flashes from a signaling society.

So one reasonable strategy change would be to extend our dwell time on each SETI target from a few minutes to, say, an hour. Doing so isn't difficult, and it could be worth the effort (and slowdown of our observing program). After all, our current SETI schemes assume that ET is either broadcasting everywhere in the galaxy at once (which, as we've noted, is a costly enterprise) or is relentlessly targeting us. We've already pointed out that it's highly unlikely that any aliens know that *Homo sapiens* is puttering around our planet: The signals that would betray our presence haven't yet reached far enough into space to alert anyone. Consequently, it's hard to imagine why they would send signals year after year to a world that, as far as they know, might have evolved nothing smarter than trilobites or triceratops. From their point of view, Earth is just another hunk of biology-encrusted rock, one of (perhaps) many billions of similar worlds in the Milky Way. The presence of *thinking* beings on Earth is, for them, only a conjecture.

WHERE SHOULD WE LOOK?

Searching for intermittent pings is surely a strategy that SETI should take under its wing. But stepping back from the matter of what sorts of signals might be headed our way, do we have any reasons to revisit our historical assumption that the aliens are camped out on planets similar to Earth?

We've always pictured ET inhabiting a world that we ourselves would find comfortable—with liquid water on the surface, a thick atmosphere, and the usual biological prerequisites. While this accords with our ideas about the conditions necessary for carbon-based life to arise and flourish, the disquieting thought remains that this presumption is narcissistic. Assuming that the aliens are on one of Earth's cousins might be just hominid hubris, akin to the tendency of husbands to think that their wives would like to be given the same sort of birthday present they want (e.g., a band saw).

One thing we can say with confidence about any extraterrestrials that are trying to get our attention is that their technical level is higher than ours. After all, they're doing the pinging, and we're not. They might be a century beyond us, or they might be hundreds of millennia ahead. In other words, at least some of these beings could be advanced enough to have emigrated off their natal worlds.

If this means no more than that they've spread throughout their own solar system, SETI researchers won't be fazed. When we examine a star system with our telescopes, we're able to pick up a signal no matter which planet it might be coming from. It's like examining a desert island from the air—any castaways will be visible no matter where they stand and wave.

But what about civilizations that are thousands of years further along than we are? Perhaps they've not only left the world of their birth but have decamped to a different star system entirely. If they've relocated to another Earthlike planet, that won't affect our search strategy either—one such world is as likely to be on our hit list as another. But there's also the possibility that a particularly expansive society may have colonized several solar systems, perhaps even retooling planets to make their environments more friendly ("terraforming" is the name given to such worldwide engineering projects). If, in the next half century, we discover a dense cluster of star systems with planets whose atmospheres are laced with oxygen, methane, or other clues to life, we might conclude that terraforming aliens were improving the neighborhood and had set up a local federation. Such a cluster of Earthlike worlds would be an obviously worthy area to sniff for signals.

If really advanced societies are scattered through the galaxy—civilizations whose space exploits make the voyages of the starship *Enterprise* look like an afternoon canoe trip—that raises other possibilities. These top-drawer aliens may be capable of constructing transmitters at special locales they know will be examined by any savvy society. As an example, pulsars—dead stars that periodically flash light and radio signals into space—are of endless interest to radio astronomers. Although there may be a few hundred thousand detectable pulsars in our galaxy, there aren't billions (the possible plenitude of Earthlike worlds). So a remote alien transmitter situated near a pulsar stands a good chance of being found by any technically capable beings. After all, pulsars are a relatively rare cosmic species.

In the same vein, a certain kind of very large, very bright category of stars—so-called Cepheid variables—might be data servers in a kind of galactic Internet. Every civilization with competent astronomers is sure to know about Cepheids—they turn out to be useful in determining the distances to galaxies, for one thing. But physicist John Learned at the University of Hawaii suggests that you could modulate the light output of one of these stars (which is to say, you could change its brightness) by firing neutrinos into its hot little center, thereby disturbing the Cepheids' nuclear fires. This is not easy to do—the neutrino gun would consume at least a billion trillion watts—and the data rate would be abysmally slow, since big stars can't flash very quickly. But the idea of using these stellar objects as smoke signals has obvious appeal, since they can be seen from far off.

Other possible high-profile transmitting sites might include black holes, which like pulsars will be studied by any scientifically literate beings. The center of the galaxy has already been mentioned as a premier location for a powerhouse transmitter, although on average it would be a great deal farther away from any aliens' home world than the nearest black hole or pulsar. Other suggestions for SETI targets have included supergiant stars—which are remarkable and rare, and therefore also of universal interest in the same way that the Cepheid variables are—and supernovae. The latter are, once again, guaranteed to be a subject of study for astronomers of whatever stripe. Clever extraterrestrials might construct their transmitters to take advantage of the fact that, shortly after one of these powerhouse explosions takes place, astronomers will be certain to have their instruments aimed toward the cataclysm. By beaming their hailing signals 180 degrees away from the supernovae, the aliens can be guaranteed a listening audience, since everyone behind them will be studying this remarkable event.

Another perennially popular suggestion is to simply forget about looking at individual targets—be they star systems or supernovae—in favor of a wholesale search of the cosmos. As a personal example of how this might be done, I joined Australian scientists Ron Ekers and Roberta Vaile in 1996 to briefly examine the Small Magellanic Cloud, the third closest galaxy to our own (it is ten times closer than Andromeda). The SMC appears to residents of the Southern Hemisphere as an unimposing, faint, and fuzzy patch on the nighttime sky. However, despite being a galactic pipsqueak, this dwarf nebula contains a few tens of billions of stars.

We used the Parkes antenna to select a hefty helping of this galactic complement—tens of millions of stars with each pointing of the telescope. The downside of such en masse examination was that all of these dense clots of stars are much farther away than those that SETI typically chooses. In the case of the Small Magellanic Cloud, which is 200,000 light-years from Earth, any signal would be about a million times weaker than if it came from a stellar neighbor in our own galaxy. In other words, to be heard across inter*galactic* space requires transmitters able to produce a signal at least a million times more intense than signaling to nearby stellar pals. It may be unrealistic to expect aliens to muster the megawatts to bridge such enormous distances, but the best way to find out is to do a few experiments. In the course of our own search, the Magellanites were mum. Or at least their broadcasts were too weak to be detected by us.*

As this abbreviated list suggests, scientists have imagined all manner of novel—and possibly more effective—celestial locales where we might run an alien signal to ground. SETI, which might seem to be a thoroughly repetitive and unrewarding experiment, is anything but—thanks in part to the unending parade of new ideas about where we might expect to find evidence of signaling technology.

The reader will undoubtedly have noticed that a majority of these ideas presuppose an interest on the part of the aliens

---

**This was not the only time SETI experimenters aimed their telescopes at other galaxies. Frank Drake and Carl Sagan used the Arecibo antenna in 1975 to search for transmissions coming from the nearby spiral M33 as well as from two dwarf galaxies, Leo I and II. No compelling signals were found.*

Stars in the Small Magellanic Cloud, a nearby galaxy

to find us, or at least to find *someone*. One of SETI's early premises—that we might blunder into a signal that was inadvertently oozing off an alien world in the same way that television, FM radio, and radar signals leak off ours—has become unfashionable of late. Everyone knows that rooftop TV antennas and set-top rabbit ears have become scarce. In much of the world, television (and the Internet) no longer reach us via spindly broadcast antennas perched on the hills outside of town. Instead, we pipe information into our homes and our lives via cables or small satellite dishes. Multikilowatt transmitters are disappearing. This development not only saves energy but also blesses us with the sublime pleasures of having hundreds of television channels and, in the case of the Internet, a two-way communication path. There's little doubt that high-powered leakage from Earth (with the possible exception of radar) will soon be a thing of the past. So if we are making this technological switch a mere century after inventing radio, you can be sure the aliens have already opted for energy efficiency and done the same. SETI scientists currently believe that any signal we find will be one deliberately beamed into space, not the Klingon equivalent of *I Love Lucy.*

If I point out these facts in dinner conversation or during a lecture, I'm nearly certain to provoke someone to ask, "Are *we* deliberately broadcasting to the aliens?" The answer to this is a barely qualified no. True, a few intentional signals have been sent skyward, but these were in the nature of demonstrations or, in some cases, mere stunts. We've never committed to a sustained transmitting effort, in the sense of using enough power and large enough antennas. My negative reply inevitably prompts the not-unreasonable follow-up: "Well, if we're not broadcasting, why do we believe that they will?"

This certainly sounds reasonable; SETI thinking, after all, is guided by the principle "Do unto them as you'd have them do unto us." But this golden rule presupposes that the players are of equal standing. Putative aliens—at least any we have hope of contacting—are bound to be far more advanced. Radio, radar, lasers—those are all recent human developments. We're just getting started. But we presume that alien societies could have developed technologies that are tens of thousands of years old or more. ET's history textbooks may not even describe the invention of radio on their planet, in the same way that our histories barely mention the invention of the wheel.

Ergo, a transmitting project may be a far less burdensome endeavor for an advanced alien civilization than for ours. "Let the aliens do the heavy lifting" is my attitude, especially as we have difficulty enough raising the funds for our receiving efforts. Not only does listening entice us with the prospect of an earlier success—we won't have to wait dozens or hundreds of years for someone to respond to a broadcast—but it also has the advantage that if we do find a signal, the prospects for transmitting become both more clearly focused and more interesting. It will be conversation, not simply a celestial shout-out.

WHAT'S THE PROTOCOL IF WE HEAR SOMETHING?

In the dusty pastures edging the town of Hat Creek, splayed beneath the northern shadow of Mount Lassen, the Allen Telescope Array is now panning the sky. This instrument, which will eventually brandish 350 antennas, is—as we've noted—destined to dramatically speed up our search for signals. Compared with earlier efforts, it will turn SETI on its metal ear. We're not talking about the difference between a Lexus and a Toyota but the difference between a Lexus and an oxcart.

In the next two dozen years, the Allen Telescope Array will make a mammoth reconnaissance of our galactic backyard, looking for signs of intelligence among the nearest million star systems. If other sentient beings populate our neighborhood, we could turn up a signal.

But then what?

This is among the most commonly asked questions of SETI: What happens in case of a detection? Conditioned by television, movies, and a penchant for expecting conspiracy, many people believe that the truth would not be out there. They think it entirely reasonable that the military will assume the aliens are a threat, surround the telescope with chain link, and redirect the data to the Pentagon. Other conspiracy theorists believe that the government will keep the discovery under wraps. The government would fear that once people hear the news, they'd lose their cool, stampede the streets, and provoke the seismic collapse of polite society. Some even venture the notion that the SETI scientists themselves would actually choose to eschew future funding and eternal fame by squirreling away their find.

Not a chance. In the case of the seductive false alarm of 1997, the scent of a possible SETI signal wafted into the offices of the *New York Times* within 12 hours of first showing up on our computers. That should be proof enough that the handful of researchers conducting SETI experiments are intensely keen to find treasure, not to bury it.

But what *would* scientists do upon receiving an incoming signal? They have nearly a half century of dealing with radio silence. So what procedures—if any—are in place for dealing with success?

In the days of the NASA SETI program, a committee drew up a document (often referred to as a protocol, although that sounds malevolent) outlining the steps to be followed by any individual or organization that finds an extraterrestrial signal. Put together by an international group of SETI scientists as well as Michael Michaud, an experienced diplomat, the document was originally intended to lessen suspicions in both the U.S. and the U.S.S.R. that a SETI detection by one or the other country might be misrepresented by, or even kept secret from, the other.

The *Declaration of Principles Following the Detection of a Signal from Extraterrestrial Intelligence,* a serious and succinct document that takes less time to read than most kitchen appliance instruction manuals, boils down to the following action plan:

First, the discoverers should verify that the signal is really extraterrestrial and artificial, not man-made interference or natural cosmic static. Having done so, those who made the discovery will notify all the other signatories to the document so that they can independently proceed to check it. They should also inform national authorities. Next on the list of those notified are all the world's astronomers, so that every available telescope can be used to study the source of the signal. Finally, the detection "should be disseminated promptly, openly, and widely through scientific channels and public media."

Sounds straightforward, right? Unfortunately, it's saddled with several serious limitations. For one, the protocol specifies behavior only for those who have signed the document, and that doesn't include all of today's SETI experimenters, let alone any other scientists who, in the course of non-SETI research, might stumble across evidence of aliens. (For example, what if future astronauts find an ancient time capsule left by visitors on the moon or Mars?) The protocol also severely limits the pool of researchers available to verify a suspected signal. In practice, these tenets of the *Declaration* would surely be ignored. But the real difficulty with this attempt to proscribe procedure is conceptual: The protocol assumes an orderly procession of events in which detection is quickly followed by verification, which is then followed by spreading the news.

That may be the way it happens in science fiction, but real life is messy. The disorder starts with the fact that the antennas used by SETI, which are coupled to digital receivers monitoring a hundred million channels or more, turn up signals all the time. In the movies, a control-room oscilloscope suddenly bounds from flat line to a profile resembling the Matterhorn, prompting stultified scientists to wake up and start screaming. In reality, sifting through all those signals for the characteristics of an *extraterrestrial* source takes a long time. Days would pass before we'd be sure, or at least reasonably sure, we'd found ET. Don't count on a press conference announcing a detection any sooner than a week after a signal first shows up on a display screen. That's an unavoidable and important fact.

In terms of dissemination, SETI is done in the open. For example, at every radio telescope we've ever used, there are observatory personnel in the control room day and night, not to mention local visitors and a raft of other interested parties (Jimi Hendrix's sister, several Silicon Valley executives, Isaac Asimov's daughter, and Miss Puerto Rico are among many who toured the control room at Arecibo during Project Phoenix observations). If we're investigating a promising signal—one that's passing tests designed to separate local interference from an extraterrestrial transmission—then a lot of people would know even before we call up someone in another state or another country to verify the signal's reality. The excitement would begin to build long before the detection was confirmed.

Because verifying a transmission is slow and the media are fast, there will be a long period of uncertainty about any newly detected candidate signal. That means you would be media blasted about a possible detection days before the people who found it were certain it was real. You can expect to read about the discovery of aliens in the checkout line of your supermarket long before it is confirmed (and depending on your reading preferences, you might already have done this.) There's also the sure-as-taxes consequence that the future will serve up many false alarms, as reporters describe interesting signals that, upon closer inspection, turn out to be telecommunication satellites, airport radar, electronic noise, or the occasional hoax.

This last possibility—that a deliberate prank might prompt a news story announcing that the aliens are on the air—is clearly more than an idle hypothesis. The claim of a signal from EQ Peg shows that. And, as we've noted, poll after poll has revealed that the majority of Americans say they believe that aliens are already *here,* buzzing our defense facilities and occasionally performing experiments on your sister. The Feds are presumed to know this, and to have the dead aliens bottled and boxed in Area 51 or some other secret warehouse facility. So a large segment of the public is primed to expect a cover-up, and is convinced that SETI signals would be kept under wraps. The need to disprove this conviction means that false alarms could become a never ending burden on the research community, forcing scientists to use scarce telescope resources to chase wild geese.

In astronomy, a situation that mimics a discovery—an event that's "close, but maybe no cigar"—can easily confuse both the public and the pros. Perhaps the most renowned of these in recent times was the matter of the meteorite ALH 84001, a hunk of Martian rock that was picked up in the Antarctic. Were there fossilized remnants of red planet residents in that meteorite, as claimed by NASA researchers, or not? In 1996, when this story hit the papers, the answer was uncertain. New discoveries are often tentative, and the public remains confused.

A more dramatic example of this type of misunderstanding happens every time an astronomer uncovers an asteroid deep in the solar system that seems to be on a collision course for terra firma. Unlike the Martian meteorite, this bit of research news could seriously impact your lifestyle by, for example, obliterating the city in which you live. Depending on the perceived threat level, you might wish to stock up on canned goods and take up residence in the subways. But how can you evaluate whether any particular chunk of incoming rubble is truly worrisome? In 1995, this problem motivated planetary scientist Richard Binzel to propose a simple ten-step scale for quantifying asteroid threats. Since his index was eventually adopted at a meeting in the northern Italian city of Turin, it's known as the Torino scale. It conveys in a simple way the level of concern as judged by the experts.

Since a SETI detection could have important societal consequences too, an international group has devised a tool to help both the public and the press evaluate any claims of extraterrestrial signals or contacts. The International Academy of Astronautics, which is a worldwide organization of professionals whose jobs are related to space (mostly spaceflight—the IAA was the idea of rocket scientist Theodore von Kármán), has a small, permanent study group devoted to SETI.* In 2000, several members of this group, including Ivan Almar of Hungary, Jill Tarter, and Paul Shuch, proposed an index inspired by the Torino scale that could be used to assess the credibility and societal import of any SETI claim.

The Rio scale, named after the Brazilian locale of the IAA's annual meeting during which this index was first discussed, ranks both the *importance* and *credibility* of pronouncements

**The author has chaired the IAA's SETI Permanent Study Group since 2003.*

that evidence for extraterrestrial intelligence has been found. The Rio scale, like its predecessor, runs from 0 (forget it) to 10 (this is definitely *it!*). The index is computed from two multiplicative factors: a term that measures the importance of the discovery (for example, if a signal comes from within the solar system, it's adjudged to be more important than one emanating from the far side of the galaxy), and a second term that estimates the credibility of the claim (is it supported by hard data, or only an anecdotal report?).

These factors are gauged by members of the SETI community. That may sound like inbreeding, and of course it is; but it's hard to think of any group better qualified to weigh in on a purported detection.

So, picture this future scenario: You log on to your computer for your daily dose of celebrity gossip and sports scores and find that someone with a backyard antenna is shouting that he picked up a signal from 47 Upsilon Andromedae. The papers don't know whether to run this story in 72-point type above the fold or bury it with the news of the weird. Fortunately, an international team of SETI cognoscenti has sifted through the evidence and given the claimed detection a Rio scale ranking, thus helping the media help the masses. As new information about the claim becomes available, the Rio ranking adjusts as appropriate.

This idea can be applied both to signal and physical artifacts. But like a foreign language or a significant other, the scale is useful only if it's familiar. In 2005, to acquaint a wider community with the index's utility, Almar and I tried applying the Rio scale to a sample of extraterrestrial detections everyone knows—the alien encounters in Hollywood films.

Consider the 1997 sci-fi blockbuster *Contact*. When Jodie Foster, monitoring cosmic static at the Very Large Array, first hears strange sounds in her earphones, her discovery ranks between 4 and 8 on the Rio scale. Once the signal is confirmed by radio telescopes elsewhere, the ranking edges up to 6 to 10. And when a transponded 1936 German TV broadcast is found buried in the signal, we've learned something even more profound: the transmission is intended for us, and it comes from only a few dozen light-years away. The final Rio scale ranking is 9.

Almar and I applied the Rio scale to a half dozen popular film and book scenarios, as well as to the EQ Peg hoax and to that endlessly contentious case of planetary physiognomy, the Face on Mars. It seemed to work on paper, and we were encouraged to promote its use more widely.

The Rio scale is not a perfect gauge; it's not magic. Unlike the Torino scale (or the moment-magnitude scale for earthquakes), the Rio scale involves factors that are hard to quantify, such as those that govern the credibility of a claim, an issue that is ultimately a matter of the expert's considered opinion. That's OK, because although experts can be wrong, they're more often right (which is what qualifies them for the sobriquet "expert" in the first place). The media and the public invariably ask what the professionals have to say regarding discoveries of potentially great consequence. Using the Rio scale would quickly have settled the matter of EQ Peg, which unfortunately lingered in the news for weeks thanks to clunky communication between the scientists and the media. Regrettably, although the scale has been written up, refined, and even tested hypothetically, it is not yet widely known among science reporters. Perhaps another hoax is needed to get it out of the laboratory and into the real world.*

**This is not intended as encouragement to the reader.*

ANSWERING ET

Although we don't deliberately send much radio traffic into the cosmos, that situation could easily change. Detecting a signal from space would strongly motivate us to pick up a microphone and start a chat, despite the tedium of waiting centuries or more for each exchange.

The possibility of sending messages was foreseen by the drafters of the *Declaration of Principles.* They also anticipated a concomitant difficulty: Who would speak for Earth, and what should they say?

I once thought that worrying about what we might broadcast to extraterrestrials made as much sense as fretting over the small talk I'd venture with Queen Elizabeth if I were knighted. I didn't dwell on the problem, as it was both hypothetical and irrelevant. Well, I've changed my mind. Not about the chances for a knighthood, but about the value in devoting some cranial calories to the matter of interstellar messaging. Doug Vakoch, a colleague of mine at the SETI Institute who has penned a number of erudite articles on the problem, has helped cause this shift. A few of his insights have percolated through the walls that separate our offices. In addition, the development of the Allen Telescope Array affords SETI the delicious possibility that we could retrieve a message from another world within a few decades. The idea of "talking back" has, in my view, become more than just an academic straw man.

Pondering what to say and how to say it could even help us snag that extraterrestrial signal in the first place. Not because it might elicit a reply, but because transmitting might give us a valuable clue as to what we're looking for. This was the view held by the author of the Project Cyclops report, Barney Oliver, late in his life (and currently endorsed by Vakoch): By confronting the technical challenge of sending signals, we'll educate ourselves in the best broadcast strategies, and can therefore shape our SETI programs accordingly. Oliver also seemed to feel that our species had a moral imperative to step up to the lectern, to do more than merely lurk and listen. As an added benefit, any aliens picking up our broadcasts would be able to ascertain our technological level, and could adjust their responses accordingly.

Sending messages isn't an entirely novel endeavor for humankind. Space buffs will recall the gold-plated plaques that were affixed to the Pioneer 10 and 11 spacecraft, launched in the 1970s. These simple greeting cards were engraved with a drawing of a friendly couple in nature's garb, together with a map of our location in the galaxy. The idea of bolting a message onto the spacecraft had originated with reporter Eric Burgess and (then reporter) Richard Hoagland, of Face on Mars fame. Shortly before the launch of the first of the Pioneers, Burgess and Hoagland were visiting TRW Systems, in California, where the craft were being tested. Looking at them, Burgess and Hoagland had a small epiphany: These would be our first emissaries to deep space. They suggested to Carl Sagan that the Pioneers should carry a message, just in case they were ever picked up by beings far beyond our solar system. Sagan thought it a good idea and collaborated with Frank Drake to design the Pioneer graphic in three weeks' time. Linda Salzman Sagan (Carl's wife) drew the artwork, and the plaque was engraved at a bowling trophy shop south of San Francisco. Today, these simple photoengravings are edging into interstellar space 7 billion miles (11 billion kilometers) from where they were etched.

The Pioneers' interplanetary successors, the Voyagers, continued the tradition of labeling our space hardware with informative material. They carried a crude videodisc with music, voices, and a small selection of inoffensive photos that could be played with a $16^2/_3$ rpm turntable and a mechanical needle (neither were supplied). Again, Sagan and Drake largely made the content selection. Voyager 1 is now the farthest man-made object in existence at a distance of 10 billion miles (16 billion kilometers), or nearly three times as far as Pluto.

The Pioneer and Voyager spacecraft are cruising toward the stars at roughly 9 miles (15 kilometers) a second. But other messages to the stars are moving at 20,000 times this speed. In 1974, Frank Drake and his colleagues used the Arecibo radar to transmit a short hello to the putative inhabitants of the globular star cluster M13. This spartan graphic amounts to only 210 bytes of data, but it has been traveling into space at the speed of light (although even at that prodigious pace it will take nearly 25,000 years to reach its audience).

A more garrulous transmission was sent in 2008 from a 230-foot radio telescope in Yevpatoria, operated by the National Space Agency of Ukraine. For some time beforehand, a website was used to collect suggested messages from the public, and in October a selection of these was beamed to the star system Gliese 581. This particular system is known to have multiple planets, including one—Gliese 581c—whose diameter is only half again as large as Earth's. This bigger brother of our home planet is also situated at the right distance from its sun to have liquid water on its surface. When discovered in 2007, Gliese 581c was touted as the most promising planet for life yet found beyond our own solar system, making it a particularly attractive target for a radio message.

Russian astronomer Alexander Zaitsev orchestrated the Ukrainian broadcast, beaming 501 personal dispatches to Gliese 581 with the telescope's 150-kilowatt transmitter. In the year 2049 we can start listening for any reaction by Gliesians to this cosmic call.

The Yevpatoria broadcast was exceptional, though, because most of our deliberate messages to aliens have been, like Igor, short and simple. I feel this has led us to imagine that any future interstellar transmissions would also have to be compact and easily intelligible to beings who are unlikely to have mastered colloquial English. Consequently, those who think about actively trying to get in touch routinely suggest that the correct language to use with aliens is music or mathematics.

But I suggest that we may have been too timid in our thinking. When Samuel Morse tapped out the first intercity telegraph message in 1844, it consisted of a paltry four words ("What hath God wrought?") That was not surprising, given that the bandwidth of the telegraph was limited, as was the patience of the crowd. The bandwidth for interstellar messaging does not have to be low, however. At microwave radio frequencies, you could easily send a megabyte a second. At infrared light wavelengths, you could up the bit rate to a gigabyte per second over long distances, and a hundred times more over shorter spans (say less than 1,000 light-years). These transmission speeds are largely set by the dispersive effects of the hot gas that fills interstellar space, and they vary a bit depending on direction and wavelength. But the point is, you need not skimp on the information you broadcast to cosmic listeners. The data pipe is fat.

For example, a society outfitted with an infrared laser of sufficient power could send the equivalent of the *Encyclopedia Britannica* to a million solar systems in a day. And each 24 hours it could beam a different message.

So this is my take on message construction. Forget about sending mathematical relationships, the value of pi, prime numbers, or the Fibonacci series. Rid your brain of the thought that aliens are best addressed with musical arpeggios, à la *Close Encounters of the Third Kind.* No, if we want to broadcast a message from Earth, I propose that we just feed the Google servers into the transmitter. Send the aliens the World Wide Web. It would take half a year or less to transmit this in the microwave; using infrared lasers shortens the transmit time to no more than two days.

Sure, the Web contains a lot of duplicate information (and a lot of adult content too, but after all, that's part of the human condition; frankly, it seems improbable that the aliens would be offended). Plus, Web pages are largely in English, which even extraterrestrial universal translators might fail to parse. But the point is, with so much redundant information, truly clever beings would have enough data to decipher the message successfully. When Jean-François Champollion decoded the Egyptian hieroglyphics in the 19th century, he benefited greatly from access to the Rosetta Stone. But even without that rocky cheat sheet, someone would eventually have puzzled out the hieroglyphics, simply because there are (literally) tons of the stuff. The same situation obtains with the Web. The profusion of data and perhaps the pictures would help any translator. In addition, given that two-way interstellar conversation will be slower than molasses in Greenland, we'd have a strong incentive to say everything at once. Any cosmic confab will be de facto one way.

So my suggestion for transmissions is to not worry overly much about the explicit content. After all, how much good would it have done Columbus to have come with prepared remarks, drafted by his sponsors, to be read to the Caribbean natives? Just send it all: the good, the bad, and the unattractive. A panoramic view of the human condition is probably the best thing we can transmit.

Despite my personal predilection for sending as much information as we can, the choice of what to say to galactic pen pals remains murky. Recognizing the problem, the *Declaration* opts to defer the issue of content, saying only that any response to a detected signal should be made only after international consultation. The idea behind this seemingly laudable thought is simple: the *Declaration*'s framers wished to forestall a situation in which one country—perhaps the one finding the signal in the first place—would monopolize interstellar conversation. Clearly, given the potential impact of such communication—imagine what might be learned from a society that is tens of thousands of years beyond our own—no nation would want to be left out in the cold—sitting by as others read the cosmic mail.

So before anyone barks back their own philosophical and political views, the *Declaration* insists that they submit their talking points to "international consultation."* But the protocol applies only to signatories, as noted earlier. Anyone who hasn't signed on to the *Declaration*—which includes approximately 99.999999 percent of the world's population—is exempted from this fastidious requirement. Indeed, at a SETI conference held in the early 1990s, physicist Freeman Dyson predicted that, in the case of a detection, everyone with the ability to bolt a transmitter onto a backyard satellite dish would soon be sending the aliens their personal credos. Dyson thought this cacophony was quintessentially human and therefore OK by him.

**The* Declaration *is thoroughly ambiguous about what this means. What is the makeup of this review body? Does submitting your broadcast script to the Swedish backgammon team qualify as international consultation?*

All of this worry about the authorization of our communications may not matter, in the same way that it didn't matter what the iguanas said to Charles Darwin when he set foot in the Galápagos. But I am compelled to note that not everyone shares my devil-may-care attitude about messaging the aliens. Some members of the SETI community—and outside it as well—think that the *Declaration* should be fortified with a stricter injunction regarding broadcasting. These folks argue that *any* transmission is potentially dangerous. That includes so-called ab initio transmissions—broadcasts made to the cosmos even before we've detected an incoming signal.

"You don't shout in the jungle" is the mantra of these fearful types "because you don't know what's out there." Their argument boils down to this: Sure, the aliens may be far away, and even peaceable. The idea of extraterrestrials motoring across light-years of distance (or sending their missile proxies) to destroy our planet sounds neither practical nor neighborly. But it's not impossible, and we can't risk the planet no matter how unlikely alien invasion may seem. So this group petitioned the IAA's SETI Permanent Study Group to amend the *Declaration* to state that *no* transmissions to other worlds (often referred to as "active SETI") be undertaken without first subjecting the idea (yet again) to international consultation.

That sounds harmless enough, insurance against the sort of calamitous scenario that Hollywood loves—where aliens pick up our sitcoms, follow the signal back to Earth, and blast our planet into a cloud of clods. But implementing this plan is not simple. Do TV broadcasters need to seek international consultation before they go on the air each day? What about our airport radars, or the radars we use to study the upper atmosphere or check out menacing asteroids? For that matter, what about the kid who steps into his backyard at night, aims a laser pointer at

Betelgeuse and taps out "What hath God wrought?" in Morse code? Should he first ask the United Nations (or maybe the Belgian Waffle Guild) to opine on the matter?

Future technology clouds the matter further. Our radio signals are just the currently fashionable method by which we might alert aliens to our presence. Later in this century we may construct large space habitats that might be discerned by technically sophisticated aliens when the habitats occasionally block light from either our world or the Sun. In the more distant future, we may build enormous solar panels, a so-called Dyson Swarm (named after Freeman Dyson, the originator of the idea), that would allow us to capture far more energy than we could ever extract from the natural resources of Earth. But these, too, would be visible to extraterrestrial astronomers. Is it reasonable to circumscribe indefinitely all activities that might signal our civilization's presence, simply because an alien society may be more advanced, and possibly dangerous? Do we wish to forever limit the activities of this generation and all succeeding generations? This is a bunker mentality, and one that I don't favor.

In any event, after two years of heated discussion the Permanent Study Group decided it was unwise and inappropriate for the SETI community to adopt such a blanket injunction against anyone who dares to aim a transmitter or laser at the sky. The *Declaration* will revert to its original purview: how to respond to a detected signal.

HOW IT WILL GO DOWN

Several of my colleagues are embarrassed by the blunders and bungles that characterized the 1997 SETI false alarm and the EQ Peg hoax, as both of these minor traumas could have been avoided. But frankly, I'm grateful they occurred. They added a modicum of real experience to the endless theorizing about what would happen in case of a signal "hit." These claimed detections were greeted by a mix of uncertainty, confusion, and aggressive interest by the media.

So we now know how a SETI success will play out. There won't be cover-ups or a government shutdown of observatories. And, unlike in the movies or the minds of protocol-minded scientists, there won't be a very orderly release of the news either. Anyone who's paying attention will realize that if a SETI experiment catches so much as the whiff of a celestial broadcast, you'll be reading about it. And since the scientists will be hemming and hawing for days about whether the signal is truly extraterrestrial, your first encounter with this story will be in the least restrained publications—Internet blogs or the checkout-line press. Shop often.

We should also expect quite a few false starts in the future—signals that either seem to be cosmic but aren't or claims that seem to be authentic but aren't. There will most likely be many cries of wolf before a true lupine shows up at the door.

And if that happens? If my sunny prognostication that we'll find the aliens within a few dozen years turns out to be true, will the public go nuts? What should we expect the real aliens to be like?

PERSISTENT PARADOX

Are we alone in the galaxy? Enrico Fermi thought so—and he was a pretty smart guy. Might he have been right?

In 1950, the famous physicist made a casual lunchtime remark that has caught the attention of every SETI researcher since. Fermi was discussing the reasonable possibility that many sophisticated societies populate the galaxy. But somewhere between one sentence and the next, Fermi's supple brain realized that this assumption had a profound implication. If there are really a lot of alien societies, then some of them may have spread out.

Fermi grasped that any civilization with a modest amount of rocket technology and an immodest amount of imperial incentive could rapidly colonize the entire galaxy. Within a few tens of millions of years, every star system could be brought under the wing of empire. Tens of millions of years may sound like a long project, but in fact it's quite short compared with the age of the galaxy, which is roughly a *thousand times* more.

Fermi immediately recognized that aliens have had more than enough time to pepper the galaxy with their presence. But he didn't see any clear indication that they're out and about. This prompted Fermi to ask what was (to him) an obvious question: "Where is everybody?" In a galaxy assumed to be filled with clever beings, why don't we see any? This dissonance is known as the Fermi paradox.

A lot of folks have given this a lot of thought. They note that the Fermi paradox is a remarkably strong argument. You can quibble about the speed of alien spacecraft, whether it's 1 percent the speed of light or 10 percent. It doesn't matter. You can argue about how long a new star colony would take to spawn colonies of its own. It still doesn't matter. Any halfway reasonable assumptions about how fast colonization could take place still end up with time scales that are enormously shorter than the age of the galaxy. It's like discussing whether Spanish ships of the 16th century could heave along at two knots or twenty. Either way, they could speedily colonize the Americas.

Scientists in and out of the SETI community have conjured up other arguments to reconcile this conflict between the idea that aliens should be everywhere and yet we see them nowhere.

One possible explanation is that interstellar travel is just too costly. Consider sending a small rocket to Alpha Centauri, for example, one that's the size of the *Mayflower.* For this modest interstellar ark to reach our nearest stellar neighbor in 50 years, you'd need about 150 billion billion joules of energy. We're not sure what aliens pay for energy, but here on Earth the going rate is about ten cents a kilowatt-hour. So the fare per Pilgrim would be $40 billion. The cost alone, in whatever currency, could deter any alien society from trying to settle distant real estate. With far less expenditure, the extraterrestrials could pursue the good life at home.

But even if the aliens can afford colonization, maybe they haven't the stamina for such a vast undertaking. Subduing the galaxy entails more than sending a ship full of restless nomads to the next star. The nomads have to settle that star and then spawn Pilgrims of their own. Those émigrés in turn have to produce yet more settlers. And so on. If each colony eventually founded two daughter settlements, then 38 generations of colonists would be required to bring the entire galaxy under control. Even the Polynesians, who swept across the western Pacific, domesticating one island after another, didn't manage this. Maybe the aliens can't either.

An alternative suggestion that would explain our apparent solitude is that the galaxy *is* settled—even to the point of urbanization—but we just happen to have the bad luck to reside in a dullsville suburb. Yet another proposal is that we've been singled out for special treatment—we are an exhibit for alien tourists or sociologists. Our world may be known to the extraterrestrials, but they observe us through a sophisticated type of one-way mirror.

Although suggested resolutions of Fermi's paradox are as plentiful as crabgrass, we still have no idea which, if any, is correct. Perhaps the universe is teeming with subtle societies we can't find, or haven't yet. Then again, maybe the explanation is simple: We're alone.

I myself have reservations about the *logic* of the paradox. Fermi made a very large extrapolation from a very local observation. I might just as well look out my patio door and pronounce that bears couldn't possibly exist because I don't see any, despite the fact that, in the history of North America, the bears have had plenty of time to shamble into my yard. That would be an erroneous conclusion.

SETI experiments at least offer the promise of relegating the Fermi paradox to the dustbin of historical curiosities by proving that other intelligence is out there. In science, speculation is essential, but experiment is definitive.

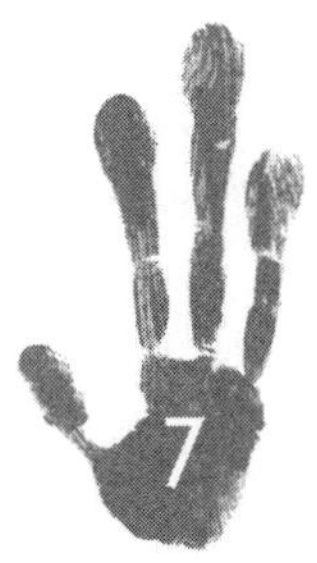

7 BEYOND GRAY AND HAIRLESS

I was standing next to a rollaway blackboard with Keanu Reeves, showing him how to write the Greek letter *mu*. Being an actor, Reeves had quickly memorized the general relativity equations I had carefully written down for his use. But his day job didn't demand the ability to scrawl formulas, so this *mu* was new. He chalked the letters slowly and deliberately, as if he were copying hieroglyphics.

We were on a set for *The Day the Earth Stood Still,* a classic sci-fi film originally shot in 1951, and now being reprised with Reeves in the role of Klaatu, the visiting extraterrestrial. My own role was rather more prosaic: The production company had flown me up to Vancouver for two days to serve as a technical adviser.

It wasn't the first time I had been asked to help with an alien movie. In 1997, when Warner Brothers was filming Carl Sagan's novel, *Contact,* several researchers at the SETI Institute were routinely queried about details of the science. Nearly every day the studio art department would call up with a request ("Can you send us some data printouts from the telescopes?") or a question. One of the latter, pertaining to Jodie Foster's high-tech ride into deep space, still sticks in my mind.

"So, Seth, what does it look like when you fly through a worm hole?" I was asked. I tried to recall the last time I had done that, but couldn't. Then I offered a measured response. "In most of the movies I've seen, worm holes are depicted by some snazzy computer animation that looks as if you're flying through a pig's intestine at high speed," I began. "But in fact, if you were to travel at nearly the speed of light, the entire universe would collapse to a bright point in front of you and another one behind. The rest of the field would be dark." That suggestion was met by profound silence. "Admittedly, that's not very visually interesting," I added, "but that's what it would look like."

"Thanks," came the reply, followed by a hang up.

In the film, this sequence looks as if you're flying at high speed through a porker's innards.

Another memorable interaction occurred one morning when the casting director showed up in my office. "I need some ideas for accurately portraying the characters in the film," she told me. "So I've come to observe your scientists and engineers." I smiled and held out my arms, gesturing to a door that gave onto a hallway full of colleagues. "Well, there they are!" I said. She left the room to hunt for scientists and engineers. After several hours she returned, triumphant. "I've learned two things about you," she declared with obvious satisfaction. "First, you all have fancy coffee mugs . . ."

This finding seemed rather inconsequential, given the effort involved. Nonetheless, my own mug did sport a SETI Institute design featuring a radio telescope and an image of how planets form. I guess that's fancy in some circles. Six months later, when *Contact* debuted, I espied an artsy coffee cup or two on-screen, so at least this bit of expensively gleaned information was apparently put to use.

Her second comment was more perplexing. "I've also noticed the way you carry your weight around here." To me, this sounded as if the casting director had somehow sensed the organizational hierarchy of the institute. As collegial as the atmosphere is, there's still a pecking order, and I figured she had perceived the ranking of the chickens. She quickly disabused me with a less insightful revelation. "You have sedentary jobs," she went on, "and so you all carry your weight around the *middle*."

I didn't contest this corporeal comment, but looked down at my waistline and compared it with those of my co-workers. In the finished movie, the technical types tend to be tubby. It was gratifying to help ensure the success of this film.

After Reeves had written *mu* several times, he turned to me. "I think I've got it." I stepped back a dozen feet, giving him some room. I waited. Next to me, the camera operator laced up one of the two Panavisions, while the director of photography made a final focus check using a tape. A grip on a ladder flagged off a light. Moments later, the assistant director shouted, "Ready, ready, ready," and the cameras rolled. "Speed," announced the operator.

Reeves moved into the shot and began writing, picking up where earthly scientist Professor Barnhardt, played by John Cleese, had already scrawled some formulas. Reeves methodically chalked one of the relativity equations, including all the subscript *mu*'s, but then added a (fictional) term that "humans still don't know." The sequence was filmed three or four times, then again from a different angle. One piece of the action disturbed me, and I sought out the director, Scott Derrickson.

"Scott," I said, "I'm concerned that Keanu's writing is a bit slow. It might not look very convincing." "Are you kidding, Seth?" Derrickson replied. "Keanu's an alien. Why do you expect him to be familiar with Greek letters?" Sounded good to me.

Mulling it over, I convinced myself that writing speed was probably a minor point. Of course, that's why I was here: to deal with minor points. I had already read the script three times, and during the past month I had suggested changes to the dialogue to make it sound more like banter between real scientists. Despite Hollywood's frequent habit, few academics address one another as "Dr. Fudnick" or "Professor Fooberg." Nor are they outfitted with clipboards, white lab coats, or a lot of Latinate vocabulary. So I rewrote verbal exchanges in the style that someone would actually use in the halls of a university astronomy or physics department.

For example, in the shooting script was this bit of supposed NASA-speak: "Object number 07/493 was first spotted outside Jupiter's orbit by the Hubble Space Telescope three days ago. It was notable for the fact that it was not moving in an asteroidal ellipse. . . . It was moving at nearly three times ten to the seventh meters per second."

I suggested that this be replaced by dialogue that was slightly less stiff: "This rock was picked up three days ago by one of the Spaceguard scopes. It's moving too goddamn fast—a tenth the speed of light!"

As far as I could tell, the production company appreciated such alterations. They were also keen to make sure that all the particulars of the set dressing were painfully authentic. At the art director's request, I had photographed bookshelves at the SETI Institute so that the right "literature" would appear in the home office of the microbe specialist played by Jennifer Connelly.

Details, in other words, including a whole lot of stuff that the cameras would never see, were important. But one point struck me as ironic: While these types of minutiae were carefully scrutinized, big picture stuff was not. For instance, the entire premise of the screenplay—namely, that aliens would come to Earth to save us from environmental self-destruction—was about as bonkers as talking to the dead. You might want to believe that aliens have some interest in our atmospheric carbon dioxide levels, but as noted earlier, none of them could possibly even know about such things. I could have pointed this out, but no one on the set of *The Day the Earth Stood Still* was going to warm to such commentary. Twentieth Century Fox had signed off on the script, and no tampering with its premise was allowed now. Somehow the cart had managed to get in front of the equine.

Still, as I watched Keanu Reeves, now Klaatu, guardedly interact with the people of Earth, I wondered for the thousandth time what *real* aliens would look like. Maybe they'd be simulacrums for tall Canadian actors, walking the streets of America and grimly warning us to behave. But somehow I didn't think so.

THE LOOK AND FEEL OF ALIENS

Nearly every audience that I address asks about extraterrestrial physiognomy. Indeed, if Earthlings ever have a conversation with extraterrestrials, I don't doubt that the first request they'll make will be "Send us a photo." We assume the aliens would want the same—witness the first-and-foremost subject of both the Pioneer plaques and the Voyager records: engravings and photographs of our appearance. "Hi, we're the lead species on Earth, and we're sending you a snap so you can recognize us at your next pangalactic cocktail party."

Of course, our interest in alien looks might be provincial, just an evolutionary inheritance from hardwired dating behavior tracing back a hundred thousand years. "What does he/she look like?" is a routine opener in the negotiations for a blind date. Personally, I suspect that curiosity about appearance—or at least bodily construction—would intrigue thinking species on any world. Nonetheless, my SETI colleagues don't often speculate on such matters. Theorizing on ET's mien is an activity they are willing to leave to Hollywood.

Unfortunately, Tinseltown might comprise the wrong crowd of theoreticians. Movie aliens are overwhelmingly anthropomorphic. The reason isn't just because the films of yesteryear (and the TV series of today) had small budgets, requiring that the extraterrestrials be played by guys in blubbery suits or latex makeup. Cineplex aliens look like us because we're programmed to quickly grasp the reactions and likely intentions of other humans, or near humans. The fact that most aliens of visual fiction look somewhat like us—upright, topped with a head, outfitted with two eyes, two arms, and a mouth that makes noise—is a requirement of storytelling, not a simple failure of imagination. Gene Roddenberry, the originator of *Star Trek,* is said to have insisted that the camera could always see the eyes of the aliens.

Of course, not all Hollywood aliens are the same. The good ones look like children, and the bad ones are reprocessed ciphers of our natural enemies—predators with big teeth or insects that can chew up the food supply. But these latter variations are the exception: The majority of the extraterrestrials on the silver screen are simply mutated versions of the hominids sitting in the audience. If such a creature walked into your neighborhood, you'd probably ignore it rather than call animal control. In reality, this degree of resemblance seems highly improbable. After all, dolphins, octopuses, birds, and chimps are all credited with intelligence, but only the last look like *Homo sapiens*. Why should the real ET bear any likeness to us at all?

Simon Conway Morris, a paleontologist at Cambridge University, thinks he may have the answer to that question. Morris suggests that convergent evolution could nudge biological development on other planets in the same directions as Earth.

"From different starting points, you end up with very similar biological solutions," Morris argues. So although the aliens might not resemble any creatures we know, their internal construction might be similar. Differences of appearance might "turn out to be skin deep. While they might look quite different, in the details of [their] organization, we'd be impressed by terrestrial similarities," he says.

We cannot know whether a real alien autopsy would reveal circulatory and nervous systems, a spiny skeleton, and a stack of specialized organs for digestion, elimination, and reproduction. The best we can do is extrapolate from the samples at hand: namely earthly life, a biota that is highly incestuous. To a surprising degree, we're all related to one another. Humans share 75 percent of their genes with pumpkins, although you might not guess this given the difference in lifestyles.

So fact number one about alien appearance is that their macroscopic structure and general demeanor is unknown and unpredictable. You should regard with suspicion the routine depictions of grays and other putative alien visitors that look much like your relatives (aside from a somewhat smaller stature, the complete absence of hair, and a distressing lack of humor).

But while most scientists would throw up their hands if you asked them to describe a real alien's appearance, they're less hesitant to opine on the extraterrestrials' biochemistry. We are carbon-based life-forms, living testament to that element's easy talent for forming stable molecules of chains and rings. Carbon spawns these well at the temperatures of liquid water, and indeed does it better than any other element. So whenever we design a spacecraft to look for life on other worlds, we naturally bias our search strategies toward carbon-based organisms. We know carbon works.

Although carbon rules in a water-based environment, does that mean it rules everywhere? How seriously should we take the oft-stated mantra that life *requires* liquid water? In 2007, the U.S. National Academy of Sciences convened a ten-person panel of biologists and geologists to consider the possibilities of what they informally referred to as "weird life"—life that's not merely a permutation of Earth's flora and fauna.

The biologists found that life might be flexible enough to gain a foothold even on worlds that you would quickly strike from your list of holiday destinations. On low-temperature planets and moons, where oceans could be filled with ammonia or methane, even carbon-based life might be able to adapt. The committee also speculated on the possibility of life in liquid nitrogen, or in the cryptic, dark oceans of Jupiter, where a hot mixture of liquid hydrogen and helium sloshes and foams. Formamide, used in the manufacture of animal glue, and hydrochloric acid have also been given the nod as possible solvents for life.

Naturally enough, the panel considered silicon-based biology. A timeworn trope of science fiction, silicon was long ago proposed as a plausible stand-in for carbon because the two elements have similar chemical properties. That may be true, but it's worth noting that despite the fact that there's more silicon in your backyard than carbon (unless you live next to a coal tip), you don't see a lot of silicon-based organisms growing or growling on Earth. Silicon had its chance on our planet and opted to beget quartz rather than bacteria. Does that reflect something truly fundamental? Is silicon incapable of spawning life?

The weird-life committee wasn't so sure. On a planet much warmer than our own, silicon compounds, particularly those in which fluorine replaces hydrogen, might flaunt their stuff. Even on worlds much colder than Earth, silicon could work—readily bonding to other atoms despite the gelid temperatures.

Nearly a half century ago, chemist and writer Isaac Asimov speculated on possible biochemistry over a range of temperatures from slightly above absolute zero to white heat. He concluded that even such extreme conditions posed no obvious barriers to the complex couplings of atoms that underpin life. The weird-life committee confirmed this expansive view, and today—thanks to exploration by spacecraft—the notion of life in unforgiving environments has taken on an even greater relevance than when Asimov was alive. In our own solar system we've discovered pools of molten sulfur on Jupiter's innermost moon, Io, and methane lakes on Saturn's satellite Titan. Indeed, the panel suggested that Titan might be a better place to look for life than Mars.

These speculations may gain greater credence in the next half century, when our robotic spacecraft plumb these unconventional worlds—and perhaps find biochemistry as we don't know it.

BUILDING BLOCKS

Despite efforts to avoid biological parochialism, those seeking life beyond Earth, whether intelligent or otherwise, still tend to narrow their reconnaissance to worlds with liquid water, where carbon compounds can hook up. We are clearly carbon-centric in our searches. We have our reasons—while all sorts of chemical systems for life might exist, carbon-based forms still look like biology's 600-pound gorilla.

Astronomers who study the so-called interstellar medium—the thin clouds of gas and dust that softly pack the dark continuums between the stars—have found roughly 150 different molecules floating in space, including such appealing stuff as water, methyl alcohol (antifreeze), acetylene, acetone, benzene, formaldehyde, formic acid, and many other items that are bottled and shelved in the back room of your high school chemistry lab. The list boasts many organic (which is to say, carbon-containing) molecules, including some sugars and a still controversial detection of the simplest amino acid, glycine (remember that amino acids are a fundamental building block of proteins). These intriguing compounds are naturally cobbled together in the relatively dense, cold knots of gas and dust that ultimately spawn new stars. That small astrophysical tidbit suggests that any solar system will be endowed with a natal supply of organic matter.

Additional evidence for the ubiquity of life's raw materials has been discovered right here on the ground. A slew of organics have been found in rocks that have plunged to Earth, most famously the Murchison meteorite that smacked into the Australian outback in 1969, about 100 miles (160 kilometers) north of Melbourne. A recent analysis of other meteorites picked off the Antarctic ice has shown that they are extremely rich in amino acids. One can only conclude

Every few weeks, a new star is born in the galaxy M51.

that when the Earth was young and subject to heavy pummeling by debris that was left over from the birth of the planets, the skies were routinely lanced by incoming rocks rich in organic compounds.

All this evidence points to an encouraging prospect: The chemical building blocks for life—carbon-based life, in any event—are made by the very interstellar clouds that eventually form solar systems. These important compounds are inevitably bequeathed to newborn planets.

Mind you, just because the organic ingredients are plentiful doesn't guarantee that life is commonplace. The Pyramids of Giza are built of limestone, but it's hardly true that everywhere you find limestone you also find ponderous, pointy edifices. However, even if a universal rain of organic material doesn't always promise life, it might bias whatever biology does arise toward carbon. This conclusion would justify Simon Conway Morris's suggestion that the aliens might be familiar once you peeled away their scaly skins. They would have their own set of proteins and other fundamental constituents, and they might not possess DNA. But carbon based? The odds are good.

These arguments underpin what you might call the conventional wisdom in our search for life on other worlds. But as reasonable as they sound, they could still fall seriously short in describing ET. Indeed, my opinion is that they do.

EXTRAPOLATING HUMANS

I've provided the arguments for why alien life might bear a biochemical resemblance to what you find in your backyard. These ruminations suggest that intelligent aliens would consist of a shambling assortment of wet amino acids, sugars, and other compounds familiar to any chemistry student. This doesn't say much about their appearance or behavior—but really, does that matter? What purpose is served by trying to envision what ET is really like, other than the amusement value in comparing such theories with sci-fi's gnarly creatures?

My colleagues often claim that they demand from the aliens no more than an ability to build radio transmitters or powerful lasers. Everything else is idle speculation about aesthetics. But despite the fact that Hollywood's aliens are often blatantly anthropomorphic, they don't always conform to the basic engineering requirements for any complex organism Consider the inventory of life's properties that you'll find in every high school biology textbook. Living things have to acquire food, process it to extract energy, and rid themselves of wastes. At least some members of each species must reproduce.

Not surprisingly, Hollywood feels little obligation to incorporate all of these features in its aliens. In 1977, I was especially discomfited by the waiflike extraterrestrials of *Close Encounters of the Third Kind.* Do they reproduce? After all, they're lacking any discernible private parts. I don't insist that their genitalia somehow resemble ours, or that they be presented for inspection. But the lack of obvious sex organs is as surprising in a living creature as would be the lack of planar surfaces on an aircraft.

Colleagues wave their hands dismissively at this nitpicking. Speculating on ET's appearance is interesting, they say, and perhaps instructive for biology majors. But it doesn't have any bearing on our SETI searches.

That may be a myopic view, because unless we can make some reasonable assumptions about the aliens' makeup, our experiments to find them could be crippled.

Consider our conception of the Martians. A century ago we imagined that the inhabitants of Mars were bipedal and accomplished. There was, we were told, an advanced and vast hydraulic society on Mars. By the 1930s our expectations had diminished, as astronomers found only a very thin carbon dioxide atmosphere on the red planet. But even in the mid-1970s we still presumed we would find plant life, or at least pond scum (if not ponds) on the Martian surface. The Viking landers failed to gratify even these modest expectations.

The disappointments continued. In 1996 scientists announced that Martians, albeit dead ones, had been found in a meteorite, ALH 84001. Alas, the claim remains highly contentious. We are left with the current, highly restrained view of Martians, assuming they exist at all: subterranean microbes.

In a century we went from Martians engaged in planetwide civil engineering to cryptic microbes a few hundred feet below the red planet's dry, pulverulent surface. Clearly, there's been a major evolution in our image of these presumptive neighbors. But the important point is that there's been a comparably dramatic evolution in our search schemes. We once used optical telescopes in Arizona to look for Martians. Now we plan the use of robotic drilling rigs.

Perhaps this change in strategy should be a salutary warning to SETI. As our perceptions of the Martians changed, so too did our techniques for hunting them down. But the search for ET hasn't been altered to nearly the same extent. Our experiments are still looking for the type of extraterrestrial that would have appealed to Percival Lowell. Admittedly, the beings we seek may not be compulsively obsessed with grooving their landscapes, but fundamentally we assume *they* are like *us*.

It behooves us to give some thought to what sorts of extraterrestrials might really be out there, producing signals we could find. How can we do that? The only path that's open is to extrapolate our own evolution.

As a starting point, consider this truism of SETI. Should we pick up a signal, it's almost guaranteed to be broadcast by a society far in advance of our own. You can easily understand why this premise must be true. If an extraterrestrial society is *less* advanced than ours, they won't be burping strong signals into space, and we won't find them. So aliens we can detect are at least as technically sophisticated as Earthlings today.

That's only a minimum requirement. Of course they could be an indefinite number of years *more* advanced.

But there's one obvious factor that might limit how far ahead of us they could be. On Earth, the development of radio technology was quickly followed by the development of nuclear weaponry. The two disciplines require similar levels of scientific expertise. So we can assume that any civilizations that are advanced enough to send a signal our way have the technology to blow themselves up. Maybe they do that routinely. If so, we won't have much chance of locating them. They'll only be "on the air" for a short period of time before they and their transmitters are obliterated by nuclear fireballs or other similarly impressive munitions.

Consequently, if we find a signal, then ex post facto we've demonstrated that technological societies (at least some of them) hang around for long periods of time—a few thousand years or more. Detection is predicated on the existence of long-lived, technologically adept intelligence. If SETI succeeds, the chance that we've found a society within, say, a century or two of our own level is clearly very small.

It's like pulling the name of a random resident of New York City out of a hat. The chance of that person being merely a month or two old is small. It's far more likely to be several years old or more.

Homo sapiens has been technologically competent for a century or so. ET will be thousands of years—at least—beyond our level. Consequently, to address the question of what the aliens will be like, we can begin by projecting a few millennia into our own future.

How could we characterize our distant descendants? According to one popular forecast, they'll brandish expanded brains, with enlarged heads to match. This is reminiscent of the grays, the iconic extraterrestrials who have bagged most alien roles in the movies of recent decades. These smoothly streamlined beings, with outsize heads set on diminutive bodies, have come to Earth to prod and probe, and their interest in us often seems disappointingly prurient. But as biologist Lori Marino once pointed out, these colorless creatures are merely predictions of what we think humans will become. The grays have no hair, minimal dentition, and only a poor olfactory sense—reflecting our own evolutionary tendencies. Their body may be small, but their eyes are big because their advanced lifestyle involves sitting in front of computers all day, writing e-mail. They are perfected humans . . . or at least, somewhat more perfect humans. Frequently, they are emotionally very cool, presumably because emotions are a defect that further Darwinian evolution will winnow from the human gene pool.*

Another scenario for our future development predicts that we'll become bionic—we'll augment protoplasm with hi-tech circuitry. The first step toward inserting engineered appliances into the body has already been taken. Often this is done to remedy physiological defects, with devices such as cochlear implants. In other cases, such as RFID (radio frequency identification) devices, embedded electronics offer some convenience (and a possible loss of privacy). But think about it: If you could encode even one bit of information into every atom of a memory device (a number that is surely too conservative), then all the information on the Internet could be kept on a device far smaller than a grain of sand. If an implanted chip could make this wealth of knowledge accessible to your brain, permitting you to surf the web while eating onion dip, then your cocktail party conversations might sparkle.

So maybe the truly intelligent aliens, the ones we could discover with our SETI experiments, are bionic. Maybe they're

---

**This is probably a conceit of sci-fi writers, particularly those penning scripts for film and television. Cognitive scientist Marvin Minsky has pointed out how useful emotions are for survival. They are thinking shortcuts that keep us alive.*

even like the Borg, with bionic bodies and a hive mentality. Plan Bee for the extraterrestrials.

But even if the aliens *are* bionic and Borg-like, that fact won't affect our experiments. We'll still be looking for carbon-based chemistry on terrestrial planets. Those assumptions are implicit in what we do now, even if we say we don't really care if they're carbon based.

COSMIC BRAINIACS

Another extrapolation of our own future leads down a different road. That's the path to artificial intelligence (AI). Its premise? We have evolved only to invent our successors. Our descendants—and presumably the aliens as well—will be synthetic and engineered, not protoplasmic and evolved.

This has become a popular theme of late. In talks I gave 15 years ago, I would suggest scenarios in which the aliens are machines, and people smiled agreeably. Today, the smiles are a bit wan. Inventor Ray Kurzweil writes about the imminent fusion of nanotechnology, genetics, and robotics to produce a new type of sentient being—one that is immortal and, in some sense, unlimited. Mathematician and sci-fi author Vernor Vinge awaits the moment when all our interconnected machines suddenly notice they have cogitating company only a data packet away, and wake up.

The imminent development of thinking machines is all the rage. Today, if you attend conventions on artificial intelligence, they have moved beyond the question of "Could we ever make a thinking machine?" to such questions as "Can we ensure that it has moral behavior?"—which is just a fancy way of saying, "Could we pull the plug, if necessary?"

However, there are people—knowledgeable people—who dispute the possibility of ever making a machine that could replace us in intellectual pursuits, a machine that could teach college chemistry, write the Great American Novel, compose symphonies, or just laze around the house watching football and drinking beer. Some folk think there's a miracle going on between our ears, despite the fact that they don't think there's anything particularly miraculous about any other parts of their body. No one rails against artificial hearts, kidneys, or other substitutes for the squishy hardware they were born with. In my opinion, to suggest that we can't construct a functional replacement for what physicist Philip Morrison called "a slow-speed computer operating in salt water"—namely, the human brain—is just whistling in the dark. We can do it. We *will* do it. The only question is when.

The AI people say they'll crank out a thinking machine within ten years. Mind you, they've been saying this for at least three decades. But, and as they point out, the lack of success shouldn't be confused with a lack of progress.

Consider the stunning implications of graphs made by futurist Hans Moravec at Carnegie Mellon University, showing the growth in computer power over the past century. You can page through endless variations of these plots, but the bottom line is invariably the same: By 2020, your desktop computer will have the computational capability of a human. It will also have a lot more memory, and it won't forget anything. Does that mean it will be able to think? That depends on software. But remember when, in the late 1990s, IBM's Big Blue computer finally clobbered Garry Kasparov at chess? Kasparov remarked, "The machine exhibited a kind of alien intelligence."

Kasparov might have been too easily impressed—after all, Big Blue was just a game machine. No one is surprised that an activity as rigidly defined as chess can be computerized. The question is, how can we move beyond a technical tour de force and build machines that actually cogitate? One straightforward approach is to imitate human brains—in other words, just knock off a proven product. Tech entrepreneur Marcos Guillen is using computers to simulate the workings of a human cortex, and his code mimics 20 billion neurons with 20 trillion connections. Other approaches start with software—sometimes in the form of interactive computer personalities, or avatars—that are let loose on the Internet and improve themselves as they respond to others. Like children, they learn. Unlike children, their mental abilities aren't limited.

None of this experimentation has yet produced a thinking machine, but the optimism from the AI community continues unabated. Kurzweil expects clever hardware by 2030. Moravec figures computerized robots will evolve to humanlike capacities by the year 2040. Tech innovator and AI specialist Peter Voss figures you can shave 10 or 20 years off even these sunny estimates. Australian researcher Hugo de Garis, who refers to these beings as artilects, urges us to bring 'em on.

"Humans should not stand in the way of a higher form of evolution. These machines are godlike. It is human destiny to create them," he says.

This sounds as if humans should throw in the towel, ceding territory to an unstoppable competitor. But I've already mentioned opponents, including several of my colleagues, who are far from convinced that solid-state hardware is ever going to replace the wetware of their crania. Barney Oliver, of Project Cyclops fame, used to quip that "artificial intelligence is real stupidity."

Indeed, considering the uninspired accomplishments of my laptop, I'm tempted to discount the notion that any future computer might metaphorically turn the keyboard around and spend its days typing instructions to *me*. Sure, computerized robots will take over routine tasks such as picking strawberries, operating subway trains, or even flying airliners across the Pacific. They've already begun replacing humans as crew in military planes, ships, or tanks—all of which are, after all, principally devices for targeting and delivering munitions. Delivery systems, including military delivery systems, will be robotized. The limitless patience and precision of machines will increasingly prove useful in operating rooms, and a robot surgeon is probably in your future.

All of these applications are straightforward, albeit slightly disheartening if you're planning on a career in neurosurgery. But could the machines ever be more than highly sophisticated tools? Machines may rival the computational capabilities of our brains, but they seem to be hopelessly outclassed when it comes to either inspiration or emotional intelligence. They may become first-rate at performing a delicate heart operation, but are we kidding ourselves if we think they will develop an appealing bedside manner?

Without the pressure of Darwinian competition or the necessity of sex to reproduce—indeed, without the simple peer pressure to get a job and establish a decent ranking in society—can we expect any machines, no matter what their IQ, to develop the complexity of interaction we find in the world of living things?

The answer to this could be yes. After all, biological complexity is not magical, nor something that only liquid-state chemistry could accomplish. It's the result of the evolutionary selection process. Admittedly, evolution is an experiment that's been running a long time. Cockroaches, for example, have been scurrying around for about 300 million years (the exact number depends on what you're willing to call a cockroach). Today's roaches have had a *half billion* generations to hone their crafty skills. It's hardly surprising that they're more adaptable than your cell phone.*

Obviously, if thinking machines depended on the slow pace of biological evolution, they'd hardly be worthy competitors to our own talents. But since they can deliberately improve themselves—because they're capable of Lamarckian evolution—their timescale for bettering themselves can be short. Whatever complexity of interaction they need to function and survive in the environment they find themselves in seems likely to emerge on timescales that will be measured in decades, not eons.

Consequently, it seems hardly unreasonable to expect that machines will soon be more than just powerful cogitators, content to spin their solid-state wheels in the service of difficult quantitative problems—they will be formidable *beings.*

I'm not trying to make the point that we might be the last generation of humans to run this planet, although that's something you may wish to think about. The real point—at least as it applies to our concept of what aliens might be like—is the dramatic discrepancy in some important timescales. Computational technology is improving at an exponential rate, with (as Intel co-founder Gordon Moore pointed out) a doubling time of 18 months. The exact numbers aren't important. What *is* important is that this rate of improvement is enormously greater than any we could expect from Darwinian evolution—even

**Considering that there have been only about 10,000 generations of* Homo sapiens, *Gregor Samsa should have been happy to wake up one morning as a far more perfected creature, namely a bug.*

making the less-than-certain assumption that the latter would move us in the direction of greater cognitive ability.

And if you're thinking "Well, we'll keep up, 'cause we're going bionic," all I can say is that you might as well believe that improved Nike running shoes will allow you to beat out a race car. If we build a machine with the intellectual capability of one human, then within 50 years, its successor will be better than all humanity *combined.*

This discussion could be regarded as merely an incentive to adopt a hedonistic lifestyle now, as the planet won't be ours for much longer. But the point for SETI is simple: Once any society invents the technology that could put them in touch—once they reach a level comparable to our own—they are only a century or two away from changing their paradigm to artificial intelligence.

The aliens—at least, any we hear—will be machines.

CONSEQUENCES FOR SETI STRATEGY

Given that ET will not be a soft and vulnerable little gray guy, but some sort of apparatus, should we in any way modify our hunt for extraterrestrial intelligence?

Well, it helps to know the habits of the prey. Allow me to list some of the advantages of sentient beings that are, in fact, synthetic:

- They would dwarf us intellectually.
- They would have indefinite life spans.
- They would be capable of Lamarckian evolution. In other words, individuals could radically improve (unlike you, despite the urgings of your mother).
- They wouldn't need all the usual accoutrements of a "biological environment."

• New machines could be fed information immediately. The young would be born smart.

One of the many sequelae of this laundry list of selling points is that these aliens can undertake interstellar travel. After all, when you're immortal—or roughly so—all trips are the same length. And these synthetic sentients needn't live on planets. What they require for existence is rather simple: some energy on which to operate and some matter with which to fashion parts, either replacements or improvements.

Does that give us any clues to where these cogitating constructions might be hanging out? Some information that could help us better direct our SETI searches? Would synthetic beings, for instance, abandon solar systems altogether, and slowly sail the dark spaces of the galaxy? Starlight throughout most of the Milky Way can provide roughly one ten-millionth of a watt per square foot of collecting area. If a thinking machine needs only a few tens of watts for its metabolism, then a solar panel several square miles in area will provide sufficient power to endlessly fire its fabricated neurons. Efficient machines could conceivably roam anywhere, not merely reside in the immediate neighborhoods of stars.

The next intuitive leap is to postulate that faster machines would eventually outclass their slower siblings. The bright bulbs would muscle aside the dimmer ones by more quickly doing whatever it is that machine intelligence does. Unlike their minimalist brethren, they would be voracious consumers of energy. A higher burn rate is required to crunch more bits per second. If you're thinking big thoughts, you need plenty of power.

That leads to the major supposition: These juiced-up, protean intellects, simply because of their level of activity, might be the easiest to detect in a SETI experiment.

If so, then it pays to ask where they would hang out. We expect the brainiacs of the universe to eschew not only the weakly lit interstellar voids, but even the neighborhoods of modest stars like the Sun. Rather, ambitious thinkers might choose to sidle up to the galaxy's true stellar hot spots. Wolf-Rayet stars, for example. There are several hundred of these superluminous objects in the Milky Way, and they boil off energy at a rate a million times greater than our Sun. That affords a considerable boost in available kilowatt-hours to any machine in the neighborhood. The galactic center, black holes, and orbiting neutron stars are other locales where advanced technology might tap into a continuous and prodigious stream of energy.

By contrast, Serbian astronomer Milan Cirkovic has suggested that the best location for cerebrating hardware would be the outer fringes of the galaxy. In those godforsaken neighborhoods, where temperatures are colder than dead penguins, energy-consuming machinery could run most efficiently. That's basic thermodynamics. But while Cirkovic's argument has its appeal, the galactic boondocks might be too dull for big brains with semi-eternity on their hands. They might prefer to exchange thermal efficiency for the opportunity to be situated closer to the galaxy's central regions, where there's a lot more astronomical action.

Do our SETI searches target any of these power-plant locales? The answer is mostly no. But perhaps it should be yes. After all, carbon-based life on Earthlike planets is so 20th century.

If, as we've suggested, the future of intellectual activity lies with synthetic brains, you might wonder what they're going to think *about*. What subjects would superior minds find worthy of contemplation? Gaining insight into their intellectual passions might help guide our SETI searches. Unfortunately, and self-evidently, it's hard to know what would stimulate an entity whose IQ is many, many times greater than our own. While sci-fi literature has ventured occasional speculation on cosmic brainiacs, I would feel certain about only two possibilities:

1. Because technological improvement would occur on a very short timescale, a "winner take all" situation would probably apply to thinking machines. That is, the first machine with the energy supplies and complexity to think fast will dominate, at least within its reach in the galaxy. The also-rans might never be able to catch up. The top dog might be the only dog that counts.

2. Space poses dangers even for machines. So some sort of selection process will inevitably occur. In order for a machine to survive, its lifestyle must preserve it against natural disaster—or possibly against deliberate disaster, if such a threat exists for machines. Perhaps successful machines make lots of copies. Or at least a few copies, which they then scatter over many light-years of distance so that one catastrophe will not destroy them all. They must do *something* to survive inevitable death at the hands of supernovae or gamma-ray bursts, among other things.

The desire of a synthetic intelligence to ensure survival by making and dispersing duplicates of itself could lead to communication over interstellar distances. After all, the many copies require updating as each acquires new knowledge or new abilities.*

**One might wonder if all the "daughter" copies of an AI entity will show eternal allegiance to the mother machine. If not, then various instances of these devices might quickly arise and effectively speciate. But again, whichever machine had the fastest improvement rate would quickly outpace the others. There would be a transcendent king that was quite unlike its subjects.*

The highly speculative scenario we've depicted here has a depressing downside. A single AI device might produce little evidence of its existence. Such a machine has no obvious reason to transmit beacons into space for the benefit of our SETI searches (other than a curiosity about antiquated protoplasmic intellects). If it is part of a *swarm* of savants, then we might possibly eavesdrop on their internal communications, should these transmissions be sufficiently energetic and widely dispersed.

But astute readers undoubtedly recognize that all such speculation about the motives, lifestyles, and distribution of synthetic intelligence is wobbly. We can no better imagine the activities and behavior of AI machinery than forest apes can divine the interests and modus vivendi of a New York stockbroker. Vernor Vinge has said that attempting to second-guess these machines is like trying to see beyond the event horizon of a black hole; it simply can't be done.

That fact hasn't dissuaded everyone. One notable effort to glimpse this AI future was made by complexity theorist James Gardner, who figured that advanced intelligence, whatever form it takes, will have a truly grand agenda. Its destiny, Gardner wrote, is to take over the universe. If these hypothesized supersentients are on track to reshape the entire cosmos for some arcane benefit, that's a project we really *could* detect. So far we've not seen the evidence. Or if we have, we've failed to recognize it.

This much seems sure: The development we've briefly sketched about the evolution from biological intelligence to synthetic sentience still lies in our future. But it surely lies in someone else's past. It is likely—I would say inevitable—that the majority of aliens beyond our technical level do not resemble the bipedal biological creatures of film and television, but have metamorphosed into a different breed entirely.

WHAT WOULD BE THE REACTION?

In the coming few decades, as new SETI telescopes relentlessly grind their way across the sky, there's both the hope and the chance that they'll find a distant summons from the cosmos that stands apart from the natural galactic noisemakers: a signal that only a transmitter could make.

Imagine that we find a signal from another world. What would be the impact?

In an informal survey I made of newspaper science writers a few years ago, every single respondent said that discovering cosmic company would be the biggest story of all time. If the "great silence" were broken, the media might go hyperbolic. However, it's less certain that the public would follow suit.

Brother Guy Consolmagno, a Vatican astronomer who writes frequently about the impact of finding alien intelligence, suggests that the people would receive the news matter-of-factly. "After all, this idea isn't new: the public has been waiting for it to happen," he says. "And they've been waiting a long time. If you look at the oldest literature—Greek mythology or the Bible—people even then were clearly comfortable with the existence of other intelligences, for example, the Nephilim in Genesis. So actually finding ET might not be a big surprise."

Is that true? Our efforts to snag signals from distant worlds are poised to lay rubber. So picture this: On an otherwise routine morning a decade or two hence, somewhere in a nondescript office building, scientists and engineers are staring at their computers when a signal from afar parades across the screens.

The word gets out, oozing into the media and people's consciousness. By the time the scientists announce a press conference, the discovery will have been trumpeted by fat headlines and analyzed by talking heads eight ways from Sunday. The truth, as well as all manner of speculation, is out there.

Would the citizenry be pleased or perturbed? Given the widespread belief in alien UFOs, perhaps they'd merely mutter, "I knew they were out there all along," and turn to the sports section.

The SETI Institute has several times tried to gauge what would happen, organizing conferences of scientists, sociologists, anthropologists, and media experts to see if they can offer insight. These pundits typically split the public's reaction into two camps. For some, the siren call from space would be enthralling. They would wish to learn as much as they could about ET, and would press scientists to send a quick reply in hopes of a two-way conversation.

This optimistic, good-to-meet-you reaction would have its opposite number. A fraction of the populace would be paranoid, figuring that humanity's status in the cosmos had just been significantly ratcheted down and we might even be in danger. Such a negative reaction might seem unrealistic since there's no physical threat—after all, the aliens won't know we've tuned them in. Plus, they're living in a star system far, far away. Just totting up the mileage between them and us would require a 16-wheel odometer. But there's a worry that ET might be sending highly advanced knowledge that could knock civilization off the rails, in the way that a modern chemistry textbook would rob a medieval alchemist of his livelihood.

Depending on the camp, reaction would range from celebration to sermonizing, from elation to agitation. People would interpret the news to fit with their own preconceived notions about aliens—for example, they're like the ubiquitous grays of TV fame. That's not so scary, though. Those ever popular, colorless aliens are pretty harmless. In addition, they're sufficiently familiar to the public that we can expect that most people won't file into the paranoid tent. Uncovering a signal might not be quite the sucker punch to our collective calm that many anticipate. This seems particularly likely since our receivers will inevitably homogenize the incoming signal, smoothing away any message as surely as a blender liquefies bananas. We'll know the aliens exist, but we won't know what they're saying.

That circumstance will disappoint the public, although it might spur investment in the far larger instrument necessary to comb the incoming transmission for content. But scientists would take solace in the fact that the discovery itself would be instructive. Biologists, for example, would have evidence that the evolution of intelligence is not a wildly improbable event. Anything that happens twice probably happens often. To the futurist, a detection will demonstrate that another life-form's experiment in intelligence has survived its own technological development, a point made by Carl Sagan. The aliens haven't self-destructed, and perhaps we won't either.

But once it sinks in that the signals are unintelligible (at least for now) and that the aliens responsible for them pose far less of a threat than those that routinely visit the multiplex, the impact of the discovery on most people will shade from the practical to the philosophical. Inevitably the public will question how important humans really are. We may be God's children, but now it will be clear that God has other offspring. For the clergy, this will necessarily force a change in outlook.

"The very learned theologians will be fascinated," says Consolmagno, "because it would provoke many interesting theological questions."

He also concurs with the widespread view that the reaction of mainstream religion would be simply to incorporate the news into existing theology. If your religion has survived millennia—if it could handle Copernicus, Galileo, and even Darwin—then ET should eventually prove palatable.

Outside the mainstream, there would be disquiet. Paul Lavrakas, a research psychologist and retired journalism professor, polled 70 members of the clergy on their reaction to a discovery of extraterrestrial intelligence. He encountered a major split in attitude that depended on the degree to which the respondent was fundamentalist. "The more fundamentalist their denomination, the less likely they were to believe in the existence of extraterrestrials."

He goes on to note, "Fundamentalist belief is extrinsic. It comes from outside the individual, for example, from scripture. It's the fundamentalist type of personality that would find the existence of aliens most threatening. After all, they'd claim that it's not in the Bible, so it must not be true."

Sociologist Bill Bainbridge agrees that fundamentalists would feel the most pain. "They'd get upset, become angry, and feel beleaguered," he says. "They'd begin to snipe at science in a broader context, and perhaps argue that this is all a hoax, similar to what we see happening with global warming. They'd argue that the signal wasn't really coming from extraterrestrials, but was Satan tempting the common man."

While such an extreme reaction might seem loony, the ensuing doubt, rejection, and attempts to distort information could be widespread. In other words, societal impact would be colored by the world situation. Consider the surprisingly fearful reaction to Orson Welles's radio dramatization of *The War of the Worlds* in 1938. That story had been around for 40 years, but on the eve of Hitler's blitzkrieg the idea of a hostile invasion had a currency with the public that simply didn't exist when the novel was first published.

Times change, and today we realize that, contrary to the *War of the Worlds* scenario, Earth will invade Mars. In the seven decades since Welles's broadcast, the general public has become far more knowledgeable about the architecture of the universe. The expanse of a cosmos in which planets could outnumber stars seems ripe with the possibility of life. Space agencies and astrobiologists actively seek microbial biology on the nearby worlds of our solar system. The consequence is a widespread expectation that other sentient life shares our universe. An experiment that proves this true might be akin to Roald Amundsen's forcing of the Northwest Passage at the beginning of the 20th century. For 300 years, people believed that a water route existed in the ice-choked Canadian Arctic. Consequently, it wasn't exactly a bombshell when someone finally sailed through it.

Historian Michael Davis agrees that the public will be thoroughly able to handle the news of an alien signal. Indeed, he believes that the most shocking thing about finding ET's broadcast could be the absence of shock, and that a majority of the population won't be especially moved.

"Having seen all those sci-fi movies in my youth, I want to believe that this would be a transformational event: bringing world peace, and giving humanity a more profound vision," explains Davis. But the evidence argues against such belief. "We've already done that experiment. We did it over a century ago." The "experiment" Davis refers to was the claim by

Percival Lowell that hundreds of interlocking canals slotted the Martian landscape, clearly the work of technically savvy beings. "The Edwardian public broadly believed this.

"I don't think there's much evidence that [Lowell's] claims had the kind of profound transformities that some might have hoped for," Davis concludes. "They seem to have left no profound cultural imprint." And they certainly didn't bring world peace.

So if persistent claims of neighbors no farther away than Mars failed to improve the public's behavior, or even get much of their attention, does that suggest that a SETI detection would be received with a yawn and a flurry of late-night comedy monologues? That may be too simple a conclusion, because it mixes short-term and long-term effects. Lowell was influential in his time, but two generations later space probes proved that the Martian canals were figments of his brain and eye, thereby defusing the lasting import of his claims.

In contrast, a SETI signal will provide data that can be verified by anyone who has suitable equipment. Unlike the Martian canals, a claimed signal is unlikely to be disproved by subsequent generations. So the degree to which a SETI success will mark a revolution in our self-image largely hinges on how many other transmitting worlds we can find, and our eventual success in deciphering any content.

A message—a decoded message—could lead us down unforeseeable paths. We've already spoken of the possibility of receiving highly sophisticated information, a disruptive event that might force us to deal with knowledge that otherwise would take us millennia to learn. Perhaps we'll be invited to join an interstellar culture network, a sort of galactic club. Even if our cherished religious beliefs could incorporate the idea of galactic brothers, what if we get ET's take on God? Do the aliens believe? Even if they're just cogitating machines, do they have insight into the purpose of existence that we don't? Soon after Captain James Cook sailed into the ports of South Sea island communities in the 18th century, those cultures abandoned their own religious practices in favor of Christianity. Cook wasn't doing missionary work, but he hove into view with impressive technology: metal, guns, the wheel, and—of course—a big ship. From the natives' point of view, whatever juju Cook had was clearly more powerful than their own. If we could ascertain that aliens also have a religion, would we be similarly seduced? Would we be tempted to join a galactic church?

Although the long-term possibilities of finding ET could unfold in many directions, the short-term reaction is more certain. Contrary to those who believe the government would squelch news of an alien signal, the past history of SETI suggests that a conspiracy of silence is neither possible nor necessary. You'll get the news that a signal has been found. You'll also learn a small number of immediately discernible facts: How far away are they? What sort of star system is it? The biggest part of this story will simply be that a signal has been received. And if the pundits are to be believed, you won't be upset to learn that someone is out there. Indeed, maybe you *will* just turn to the sports section.

When, years ago, Chinese Premier Chou En-lai was asked his opinion about the French Revolution, his reply was "It's too soon to tell." No matter how big or small the first headlines, the discovery of aliens in the deep tracts of the galaxy will be merely a beginning—a first step down a long path.

THE MEANING OF IT ALL

Now and again someone who doesn't know me very well will ask, why do this? Why choose an occupation so odd and seemingly quixotic as looking for aliens? After all, I occasionally encounter highly respected scientists who tell me that they think SETI is a complete waste of time. The evolution of intelligence is a fluke, they say. Or Earth is exceptional. Or our current techniques for finding someone out there are either too insensitive or too provincial. Many scientifically literate critics point to the Fermi paradox, or some variant thereof, and pronounce that the verdict is already in: Our galaxy is, at best, only sparsely inhabited. Even those who accept SETI's mission frequently opine that it may take generations to succeed, if success occurs at all.

So again, why would I spend my one-and-only brief moment on the stage of life chasing this particular rainbow?

I don't do it for the pay, or for the health insurance. The SETI Institute is a wonderful place—with an enticing ambience and an upbeat environment that I've seldom encountered anywhere. But that's not enough to seduce me into spending a career hoping for signals from beings whose very existence is speculative.

What *is* enough is the fact that SETI addresses a truly big-picture question. This is exploration on the scale of those European sailors who plied and plotted the world at the start of the 16th century. Unless our concept of the cosmos is gravely in error, SETI is the beginning of the last major foray into the unknown.

That's a source of personal satisfaction. Most people worry about profit and loss, advancing their career, or simply providing for their family. What a privilege to be able to address one of the truly big questions—a question whose relevance and appeal apply not just to one or two generations but to humanity's entire parade.

Am I too early? Is today's SETI destined to fail and be regarded—a century hence—as a quaint and naive idea that fell victim to an unknown, dead-obvious fact of the universe? Maybe we'll learn that solar systems like our own are as rare as albino crows. Recent computer simulations made by researchers at Northwestern University suggest that possibility—that very few planetary assemblages will resemble the orderly structure of our own, with nice rocky worlds on the inner tracks and dead gas giants on the outer.

Then again, maybe we'll fail to find even microbial life nearby, suggesting that biology is not pandemic, but merely an isolated outbreak of weird chemistry that started on our planet but on few (if any) others.

Or maybe the zoologists who claim that the evolution of intelligent creatures is a singular development will turn out to be right. In a galaxy rife with life, we could be at the top of the IQ charts, a dismaying fact that would guarantee that our SETI experiments will be barren.

Or maybe, as physicist Paul Davies has suggested, we've boarded the bus too early. Given the difficulty (and possible danger) of sending hailing signals to unknown worlds, maybe the aliens will wait until they've heard from us before getting in touch. In other words, they might not fire up their transmitters until the first of our TV or radar signals have reached their planet. At that point they'll know that there's someone here to enjoy their broadcasts and that our level of technology is low enough to be thoroughly nonthreatening. If so, we should begin our SETI experiments in the 40th century, and meanwhile get on with promoting world peace or buffing up our Sudoku skills.

Any of these arguments might have merit. But we don't know that. Consequently, the choice is simple: We can sit on our hands and wait or we can use the equipment we possess today. Moore's law makes every year a new adventure for SETI. This observing season won't just be better than last season—it will be as good as all previous seasons put together. Moving forward seems like the right thing to do because if *Homo sapiens* is the sole locus of thought in the cosmos, what has transpired on Earth is stupefyingly improbable. That conflicts with a tenet among astronomers that imagining you are special is nearly always a failure of imagination.

Yes, it's a very long yellow brick road, but we're finally taking some first steps: one that began in Oz and may yet reach a destination far more wondrous than an Emerald City. If we succeed, our descendants will look back upon these times and say, "That's when the door opened."

WILL OUR FUTURE BE LIKE STAR TREK?

It's the 23rd century, and our descendants have an odd daily drill: They suit up in spandex unitards and boldly go where no man has gone before. Thanks to warp drive, the "going" part seldom takes very long, and the gracile beings who crew the starship *Enterprise* and its endless replacements—ciphers for our great-great-great grandkids—spend most of their waking hours interacting with bumpy-faced aliens.

Welcome to *Star Trek*'s view of the future, a time when today's grueling treadmill of competitive commerce has yielded to a more academic lifestyle. Our heirs are destined to become anthropologists specializing in aliens. The ability to quickly traverse interstellar distances will open the galaxy to quick and easy exploration, and turn our lucky progeny into facsimiles of Captain Cook, but without the scurvy.

Is this really anything like humanity's future?

The *Star Trek* stories envision that we will join with other worlds in a Federation of planets, a loose assembly of more than a hundred worlds in which humans (you'll be gratified to know) are the first among equals.

It's a nice idea. But as a forecast of our future, it's no better than Buck Rogers or a consultation with a Magic 8 Ball. The problem—even aside from the quaint anthropocentrism—is the failure to acknowledge both physics and predictable technological progress. Critical to the existence of the Federation are the abilities to travel and communicate at faster-than-light speed—and we (or they) may never be able to do that. You might argue that conventional physics, namely special relativity, offers the possibility of cruising to the stars, because the faster you go, the more slowly you age. If you could zip along at very nearly the speed of light—no small accomplishment—you might cover the distance from one part of the galaxy to another in a matter of days or even minutes as gauged by your watch.

But an appeal to Einstein won't solve the problems of our *Star Trek* future. Even setting aside the enormous energy requirements of relativistic rockets, the peculiar fact that every spacecraft's clock would run at its own speed would create freakish situations. For example, imagine that all spacecraft regularly synchronize their clocks via radio pulses from Federation headquarters. That would let them know today's "stardate" when filling out the ship's log. But this won't lift the curse of clock-rate anarchy. Imagine a scenario in which Captain James Kirk transfers the helm to Jean Luc Picard, and then departs for a vacation cruise on a very high-speed rocket. He could conceivably return two weeks later (by Kirk's reckoning), in time to attend Jean Luc's retirement party 40 years later (by Picard's reckoning). Perhaps Kirk would be asked to reassume the helm of the *Enterprise*. The ship would undoubtedly have undergone a half dozen retrofits in the interim, and Kirk's training and experience would be of mild academic interest only. It's odd when you can be younger than your children's children.

This shuffling of generations is hardly the most troublesome difficulty in the Federation. Imagine an invasion of a friendly star system by the evil Cardassian union. The aggression prompts an SOS to Starfleet command. In an optimistic case, Starfleet headquarters might be only a few hundred light-years away. The duty officer radios the Federation starship nearest the besieged supplicants, one that's—say—a mere ten stars away, or 50 light-years. The Starfleet craft arrives on the scene of trouble in ten minutes by its clocks, thanks to relativistic time dilation. But hundreds of years have transpired at the star system in question, and the Cardassians have done whatever they wished to do. A direct SOS to the starship would speed up the rescue, but (in this case) a century would still have transpired between first notice of the threat and the arrival of the Federation cavalry, at least according to the clocks of those being invaded. That's not a rescue, it's a visit to a historic battlefield.

The bottom line is that the only way to hold together an interstellar federation is by using physics that is still speculative—that allows both faster-than-light travel and communication.

You might wish to assume that both will eventually be possible, and doubts to the contrary are merely due to the inadequacy of my brain. But other aspects of the *Star Trek* future are likely to prove as batty as Carlsbad Caverns. To begin with, all the members of the United Federation of Planets are sentients on a nearly equal technological and cultural level. The same applies to the surrounding Klingon, Cardassian, Breen, Romulan, and Tholian empires. That's why these various species can be credible allies or enemies. If the Romulans' intellectual rank was comparable to that of harbor seals, they would figure less prominently in the tales of the starship *Enterprise*.

But the Federation is, according to the *Star Trek Encyclopedia,* 10,000 light-years across, which means that it encompasses roughly one percent of the galaxy's real estate. Within this restricted neighborhood are a few hundred different species willing to join up with the Federation, and all seem to be within a century or so of one another's evolutionary level. Simple arithmetic shows that this implies a rate of appearance of sentient societies in the Milky Way of hundreds per year, or trillions over the course of the galaxy's history. That fecundity even bunnies would envy. By contrast, in our discussion of the Drake equation (see page 106) we pegged the production rate of intelligent societies as one or two a year.

However, the most startling assumption about this hypothesized hodgepodge of talented creatures (other than their striking anthropomorphism) is that . . . they're creatures! The Klingons, Vulcans, etcetera are still biological. The transition to artificial intelligence, a step that seems likely to be taken within a century here on Earth, hasn't happened either to them or to us, according to *Star Trek.*

This is first-drawer provincialism, and depending on your personal philosophies, just whistling in the dark. Earlier generations of Earthlings once speculated that space travel would be powered by waxen wings, harnessed swans, or black gunpowder in a giant cannon. *Star Trek*'s vision of our future—despite its extensive appeal—will undoubtedly turn out to be just as naive.

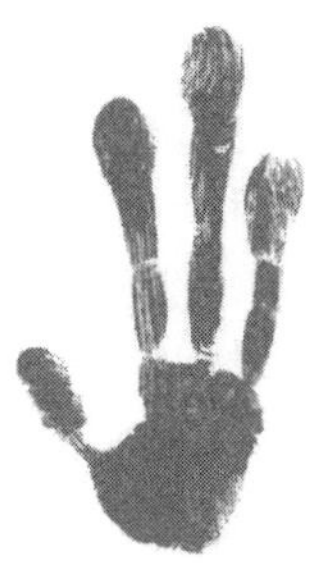

EPİLOGUE

In the last weeks of 2007, the United Nations General Assembly proclaimed 2009 as the International Year of Astronomy. This pleasant-sounding initiative was taken at the urging of the International Astronomical Union and the UN's own cultural and educational arm, UNESCO. The year marks the 400th anniversary of Galileo's first telescopic observations, and the UN wanted to promote greater public interest in science by showcasing one of its most alluring disciplines: astronomy.

The year 2009 also marks SETI's passage into its second half century. It's been five decades since Cocconi and Morrison's article in *Nature:* the seminal paper laying out the impetus and schema for finding other beings on other worlds. The SETI Institute celebrates its 25th anniversary in 2009.

We'll have plenty of opportunity for star parties, planetarium shows, and TV specials. Yet despite a near lifetime of searching for our alter egos among the stars, we haven't found any.

In 1982, science fiction author David Brin wrote of "the Great Silence," noting the failure to find evidence of extraterrestrials.

Since then the silence has deepened, prompting some to tentatively conclude that no one is out there.

Peter Ward, the paleontologist who in 2000 co-wrote the popular book *Rare Earth,* says that SETI, while an "interesting intellectual endeavor . . . probably won't find anything." Our planet is inordinately special, according to Ward, and few other worlds will be suitably outfitted by nature to spawn intelligence.

Author Michael Crichton opined in a 2003 lecture at Caltech, "SETI is not science . . . [it's] unquestionably a religion." Stanton Friedman, of Roswell fame and a gentleman whom I've debated both on radio and television, argues that the aliens are here, but disparages efforts to find them *out there.* He derides SETI as the "Silly Effort To Investigate" and refers to its practitioners (including me) as "cultists."

When SETI scientists respond to the skeptics by noting how little of the sky has been deeply searched, some say we're merely making excuses to keep working, concocting fresh tactics and building expensive new instruments. Meanwhile, we continue to forecast that success is just around the corner. Our persistence in the face of endless null results is regarded as futile stubbornness, or maybe just a symptom of mental infirmity.

Such commentary, coupled with the difficulty in raising money for SETI, seems like good reason to be discouraged. We could abandon the effort and seek cover for retreat by announcing that our experiment is premature. We really should wait for telescopes that can simultaneously observe the whole sky and monitor the entire radio spectrum. Alternatively, we could acknowledge that maybe the extraterrestrials won't signal in our direction until they've heard from us. If so, we shouldn't expect to detect anything for another few centuries. Or we could always take refuge in the unsettling premise that we've chosen the wrong fork in the technological road and are uselessly hunting for radio or light signals when the aliens have moved on to better modes of communication.

Despite all the reasons for quitting, doing so would be ill-advised. The sort of arguments the skeptics make would dissuade *any* explorer. Imagine being a member of the Spanish court in the late 15th century, counseling Columbus. You might suggest he give wooden ships a pass and hang fire for 500 years, after which he could cross the Atlantic in hours, eating low-grade meals off his lap. Yet Columbus discovered an extraordinary new world, and his wooden ships were (just) good enough to find it.

Likewise, SETI's vehicles for exploration may be good enough now, and they're improving rapidly. The Allen Telescope Array will increase the number of carefully examined stars by a factor of a thousand in the next two dozen years. In addition, a worldwide effort is under way to build a follow-on instrument, the Square Kilometer Array (SKA), by the year 2020. This impressive device will have a collecting area approximately ten times greater than Arecibo's monstrous mirror. Its frequency range (100 MHz to 25 GHz) will be more than double that of the Allen Array. While not specifically designed for SETI, this next-generation telescope—planned for construction in either South Africa or Australia—could be used in a commensal hunt for artificial signals. SETI on the SKA would have truly unprecedented sensitivity: It could find emissions from a similarly sized transmitting antenna 1,000 light-years away if the aliens could muster a feeble five kilowatts of broadcast power (similar to that of a small AM radio station).

The Netherlands and Germany are jointly constructing a radio telescope that will complement both the Allen Array and the SKA by working at frequencies at the bottom of the dial, from 10 to 240 Mhz. This instrument, known as LOFAR (Low Frequency Array), is, like the SKA, being built for astronomy, not SETI. But it will be the first very large telescope able to monitor the same frequencies as terrestrial FM radio and television.

This tedious enumeration of astronomical instruments shows you what's coming down the pike, but every SETI researcher's dream is a telescope that can observe the entire sky at once over a very wide swath of the radio dial. We know how to build such a dream device but are hindered from doing so by the requisite—mammoth—computing requirements. Using today's digital technology, even a modest omnidirectional antenna, would consume megawatts of power and cost megadollars to build. In the current funding climate for SETI, the idea is a nonstarter. But these are problems that the march of technology will eventually trample into irrelevance.

Optical telescopes are also on a rapid growth curve. California's Palomar 200-inch telescope was the alpha male of optical telescopes for much of my youth: it was the biggest eye on the sky. Today, the twin Keck telescopes in Hawaii have mirrors 30 feet in diameter, and telescopes on the drawing boards will have apertures the size of a football field.

To me, the most interesting thing about these new instruments is that they might find signals of a type that we currently don't anticipate or simply can't see: for example, a once-a-year, omnidirectional flash from a beacon that's below the sensitivity of today's mirrors.

PROVINCIALISM?

Of course, topflight instruments alone aren't going to find ET. Discovery depends on people and projects. The alert reader will note that I've discussed SETI experiments in the United States only. That's less a matter of chauvinism than a reflection of the facts. Today, the only professional, non-American experiment to find extraterrestrial intelligence is being carried out at the University of Bologna, Italy.

This lack of international interest is significant. A decade ago, I gave a talk on SETI at the University of Groningen, where I once worked. The colloquium hall for this talk was SRO. I began by asking how many in the audience believed that other worlds might house intelligent life. Nearly everyone raised their hand.

I then asked how many were willing to spend one Dutch guilder a year—adequate to buy a small cup of bad university coffee—to conduct a search. Not a guilder a day, but a guilder a *year*. All the hands went down.

After the talk, I asked one of the professors why this was so. The Dutch, after all, have the equipment, the expertise, and the money to mount a SETI experiment. Why were they so reluctant? "We're too serious for that," was his reply.

Other nationalities besides the Dutch have the ability (and funds) to do SETI, but choose not to. The British, French, Germans, and Japanese, as well as several smaller countries, all possess the hardware, wit, and wherewithal. So why have Americans been left nearly alone in the field? Does our frontier mentality, the willingness to gamble on a long shot with a big payoff, encourage us to try what other countries won't?

I used to ponder the significance of this difference (without much gratifying result), but my efforts may soon become irrelevant. Attitudes are changing. At recent conferences in Europe I have noted a new excitement for SETI. The French, Dutch, British, and Koreans are all talking about plans for doing experiments. Perhaps the steady drip of new planet discoveries has finally prodded these other communities to feel that searching for aliens is no longer just a nice premise for a blockbuster movie.

Maybe a new generation, raised on the numerous sci-fi stories that permeate television and the movies, is more kindly disposed to both the existence of aliens and our efforts to find them. This is certainly the impression I get when I lecture at universities. The undergraduates are keen to talk about my work, and the grad students want to know how they can get a job with SETI.

In addition, there's the allure of what we might gain from contact. Earlier, I noted Carl Sagan's observation that finding someone more advanced than ourselves would demonstrate that self-destruction is not *Homo sapiens'* inevitable fate. Sagan may have been overly optimistic. Just because we find one viable society doesn't rule out the somber possibility that 10,000 others pushed the button. Some SETI researchers have touted the benefits we might reap from the extraterrestrials' highly advanced knowledge. Besides learning all the physics we don't know, we might be taught the secret of immortality, or at least lessons in how to get along. While all of this is surely tantalizing, it's also contingent on the willingness of other societies to share, and our ability to understand what they're saying. So while SETI could greatly enrich human knowledge, that possibility is highly speculative.

SETI researchers often justify their work by exclaiming, "The question we're trying to answer is one that humanity has asked since the dawn of our species," as if that circumstance somehow validates the search. It doesn't. Many questions that people have asked forever are relatively unimportant (e.g., why do fools fall in love?).

Despite all this, it's hard to escape the feeling that there is something profoundly worthwhile in discovering the existence of other intelligence, whether we understand it or not. Knowing someone's out there would calibrate our position and our relevance. Either we're part of an extended family or we're alone. In addition, looking for intelligent life has an appeal that is viscerally greater than the more generalized search for biology beyond Earth. Microbes on Mars are thought provoking, but cogitating aliens are downright thrilling.

British science fiction author Stephen Baxter suspects that our motivation to pursue SETI is rooted in another aspect of our history. We have an innate longing for peers that once were. In the past few tens of thousands of years, we've managed to prune our branch of the hominid tree back far enough that our nearest relatives today are chimps. The Neanderthals are gone, as is *Homo floresiensis.* This situation is unlike, for example, the dolphins—of which many species still exist. Perhaps we miss having the other near-relatives with whom we shared the planet for hundreds of thousands of years. So we seek cousins on other worlds.

In the first half of the 21st century, our SETI searches will be extended to catalog a significant fraction of our galactic neighborhood. If star systems with intelligence are only one in a million, then the odds are still good that we'll find a few. But suppose we don't? Suppose a century passes without Project Ozma's promise ever paying off? Should we give up? Should we concede that the cynics are right and maybe no one really *is* out there?

That would bespeak a truth both staggering and terrifying—the universe exists for us, and for us only. Ten thousand billion billion stars burn for our eyes alone and countless planets glide in silent orbits without notice or care. While I can imagine being the only person on a Pacific island, I cannot conceive of being the sole inhabitant of a continent. The universe is simply too large for me to easily accept that only here, only on Earth, has senseless chemistry organized itself into sentient beings.

And so I continue to do SETI. It's an exceptional task, driven by a deep conviction that in the brutal, vast compass of the universe, there are others who know the secrets of nature, others who might be making their presence known, if we only have the wits and the stamina to look.

ACKNOWLEDGMENTS

Upon the completion of any book, it is customary for the author to effusively thank an extended list of people who have contributed their talents, knowledge, or encouragement to the effort. Depending on the merits of the manuscript, these contributors may or may not appreciate this recognition.

However, I hope those named below will be kindly disposed to my gratitude because, in fact, my regard for their help is both heartfelt and sincere. They have invariably been enthusiastic in sharing both their time and their wisdom.

My colleagues at the SETI Institute, in particular Frank Drake, John Billingham, Tom Pierson, Barbara Vance, Jill Tarter, Doug Vakoch, and Chris Neller, were always willing to interrupt their own work and offer insight as I queried them on questions of fact or history. Others who have helped include members of the SETI Permanent Study Group of the International Academy of Astronautics, most especially Paul Shuch and Stephen Baxter, as well as university scholars Paul Lavrakas, Kathryn Denning, Michael Davis, Joni Spigler, and Brother Guy Consolmagno.

My agent, Carol Susan Roth, set this book in motion. She has tirelessly labored on its behalf. The content editor, John Paine, demonstrated great understanding and expertise of the subject matter and my intentions, and was especially adept at straightening out the worst of my convoluted prose. Judy Klein was meticulous in enforcing consistency and accuracy in the text. Both have buttressed my faith in the value of professional editing, and this work is far better because of them.

Lisa Thomas, of National Geographic, took the initiative in securing the publication of this tome, and her support has been unflagging through every step of its production. She has been cheerful and accommodating even in the face of my relentless travel schedule and occasional sloth.

It is less than innovative to note that both my career and this project were made possible by my parents, to whom I dedicate this book. But they have offered unstinting backing and stimulus for all that I have done, and my gratitude to them will always be only a small approximation of what they deserve.

For the occasional triumph of inspiration over distraction during the course of composing this text, I am grateful to Nikolai Rimsky-Korsakov.

Finally, I want to thank my wife, Karen, who suffered many months of inattention while I pursued this work, and who was willing to give me at any time, and for any length of time, the benefit of her remarkable talents. If the narrative within is comprehensible and clear, that is largely because of the long hours she spent reviewing the text. That it now leaves the confines of our den and sees the light of day is due to her infinite understanding and unstinting devotion.

ILLUSTRATION CREDITS

2, Seth Shostak; 10, Seth Shostak; 22, Bettmann/CORBIS; 31, Courtesy Lowell Observatory Archives; 37, NASA; 46, Seth Shostak; 70, ESA--C. Carreau; 86, Ralph White/CORBIS; 94, Seth Shostak; 105, Seth Shostak; 118, Bettmann/CORBIS; 131, Roger Ressmeyer/NASA/CORBIS; 159, Seth Shostak; 169, NASA; 182, Seth Shostak; 189, Seth Shostak; 228, NASA; 261, NASA/ESA.

INDEX